Nonlinear Mechanics

A Supplement to Theoretical Mechanics of Particles and Continua

Alexander L. Fetter

Departments of Physics and Applied Physics
Stanford University

John Dirk Walecka

Department of Physics
College of William and Mary

Dover Publications, Inc.
Mineola, New York

Bibliographical Note

Nonlinear Mechanics: A Supplement to Theoretical Mechanics of Particles and Continua is a new work, first published by Dover Publications, Inc., in 2006.

Library of Congress Cataloging-in-Publication Data

Fetter, Alexander L., 1937–
    Nonlinear mechanics : a supplement to Theoretical mechanics of particles and continua / Alexander L. Fetter, John Dirk Walecka.
       p. cm.
    Includes bibliographical references and index.
    ISBN 0-486-45031-7 (pbk.)
    1. Continuum mechanics. 2. Field theory (Physics). I. Fetter, Alexander L., 1937–   II. Walecka, John Dirk, 1932–   III. Theoretical mechanics of particles and continua.

QA808.2.F47 2003 Suppl.
531—dc22

                                      2006040221

Manufactured in the United States by Courier Corporation
45031702
www.doverpublications.com

PREFACE

When Dover Publications agreed to reprint our textbook *Theoretical Mechanics of Particles and Continua*, we decided to write a short additional volume on *Nonlinear Mechanics* that would include material that each of us has used in recent graduate classes. This supplement focuses on various aspects of nonlinear phenomena in both fluids and discrete systems of particles. We start with the onset of hydrodynamic instabilities, particularly the Rayleigh-Bénard convective instability in a fluid heated from below. The associated nonlinear Lorenz equations provide a simple model for the chaotic behavior far beyond the onset of convection. We then study the logistic map, which illustrates period doubling as a different route to chaotic dynamics. Finally we consider hamiltonian systems, where the familiar action-angle variables provide a convenient and flexible tool. Two solvable models form the basis for a qualitative discussion of the celebrated Kolmogorov-Arnold-Moser (KAM) theorem on the stability of a separable hamiltonian system subject to periodic perturbations. In these examples, we rely on the general aspects of nonlinear dynamical systems such as fixed points and their stability. In addition, various numerical methods play an important role, especially the Poincaré surface of section. A set of problems has been included, some for each section. We are most grateful to M. E. Peskin for allowing us to use his unpublished lecture notes on hamiltonian systems. Throughout this work, we have very much enjoyed developing the material and hope that students will find it a valuable addition to our earlier textbook.

Alexander L. Fetter
John Dirk Walecka
October 1, 2005

CONTENTS

Part I
Introduction

1 Motivation

The book *Theoretical Mechanics of Particles and Continua* was originally published by McGraw-Hill Book Co. in 1980 [Fe80a]. Subsequently, Dover Publications reprinted it in 2003 [Fe03]. The original preface to [Fe80a] states:

> *We intend this frankly as a textbook and aim to provide a lucid and selfcontained account of classical mechanics, together with appropriate mathematical methods.*

Over the years, many colleagues and students have told us how much they liked using this text.

The first section of [Fe80a] starts with the sentence:

> *Although Newton's laws of motion are easily stated, their full implications involve subtle and complicated nonlinear phenomena that remain only partially explored.*

Since 1980 the advent of powerful inexpensive computers has revolutionized this exploration. Currently anyone with a desktop computer can simply pick appropriate initial conditions and numerically integrate a set of nonlinear ordinary differential equations, or, equivalently, iterate a set of nonlinear finite-difference equations. These numerical investigations have discovered many fascinating and unexpected phenomena, such as *chaos* and *fractals* (see, for example, [Za85, Gu90, Mc94, Ot02]). Simultaneously, powerful mathematical methods have been developed to describe nonlinear mechanics (see, for example, [Ar89, Li92, Pe92, Jo98]).

The preface to [Fe03] further states:

> *We have each taught particle and continuum mechanics many times over the years, both at Stanford and at William and Mary, and enjoyed having this book available as a text. ... In the past several times that we have taught the course, each of us has supplemented this material with additional lectures on more modern topics such as nonlinear dynamics, the Lorenz equations, and chaos. We hope that this supplemental material will also be available in published form at some point.*

When Dover reprinted the original version [Fe03], the authors considered preparing a revised second edition but decided that it was more valuable to have the text immediately available for the use of students and instructors. Thus arose the idea for a *supplement* that would provide a bridge from [Fe80a] to contemporary (typically nonlinear) mechanics. We re-emphasize that this material serves as a *textbook* from which one can learn. Indeed, we claim no originality and are definitely not experts on these topics.

The second half of [Fe80a] focuses on continuum mechanics with chapters on *Sound waves in fluids*, *Surface waves on fluids*, *Heat conduction*, and *Viscous fluids*. A natural extension is to use this material as a basis for discussing nonlinear continuous systems, which we proceed to do in Part II of this supplement.

The Euler equation for an ideal incompressible fluid simplifies considerably for irrotational flow because the velocity field (a vector) is then derivable from a scalar velocity potential. In this case, the Euler equation can be integrated to yield Bernoulli's equation. Part II starts with a linearized stability analysis describing two classic physical problems: the onsets of (1) the Rayleigh-Taylor gravitational instability for two fluids with the heavier on top, and (2) the Kelvin-Helmholtz shear instability where the fluids are gravitationally stable but undergo relative transverse motion. This material provides a nice introduction, for it simply amplifies three problems appearing in the original text [Fe80b].[1]

The Navier-Stokes equation adds viscosity to the description of these fluids. Typically, a viscous fluid undergoes rotational motion with nonzero vorticity (the curl of the velocity). In addition, the vorticity itself diffuses at a rate determined by the kinematic viscosity. The Navier-Stokes equation is solved for some simple physical configurations in [Fe80a]. Inclusion of heat flow, both conduction and convection, leads to still richer physical phenomena. If the fluid is heated from below, the decrease in density associated with thermal expansion substantially affects the dynamics, and the resulting buoyant force eventually initiates convection. This convective instability of a viscous fluid heated from below is known as the Rayleigh-Bénard problem. The simplest approximation is to retain only the leading linear temperature dependence in the density (known as the Boussinesq approximation). We analyze the resulting set of coupled nonlinear dynamical equations in some detail, obtaining the conditions for the onset of the striking convective roll instability and deducing properties of the linearized solutions (see, for example, [Ch81, La87, Bo00]).

The coupled physical amplitudes that obey the nonlinear Boussinesq equations can be expanded in a complete set of spatial normal modes [Sa62]. If the resulting system of coupled nonlinear equations is truncated to retain only the first two modes, then an appropriate redefinition of variables yields the equations first derived by Lorenz as a model for weather [Lo63, Sp82]. These celebrated and remarkable Lorenz equations constitute a discrete dynamical system with three dependent variables and one control parameter. As we shall see explicitly, their solution mimics the much more complicated Rayleigh-Bénard problem with an infinite number of degrees of freedom. The numerical solution to these three coupled, first-order, nonlinear, ordinary differential equations provided one of the first observations of the phenomenon of *chaos*. Today, it is a simple matter for students to solve these equations on a desktop computer and investigate their properties. Indeed, this system exemplifies much of what makes modern mechanics so enjoyable and fascinating.

For pedagogical reasons, we take an extended path to the Lorenz equations, exploring some interesting physical phenomena along the way. Part II concludes with a direct derivation based on a simplified physical situation where the low-lying modes indeed decouple and the relevant motion of the fluid is readily observed [Yo85]. Properties of the solutions to the Lorenz equations are analyzed in detail in Part III.

---

[1]Note to the dedicated reader: we here provide the solutions to those problems.

3

The first half of [Fe80a] deals with particle mechanics, and Part III of this supplement returns to the study of systems with a finite number of degrees of freedom. It begins by introducing the Duffing oscillator. This typical and important nonlinear oscillator augments the familiar quadratic harmonic oscillator potential energy with a quartic term. It exhibits very characteristic behavior (spontaneous symmetry breaking and bifurcation) as the sign of the quadratic term passes through zero (note that this behavior goes well beyond the usual stable harmonic oscillator).

The Duffing oscillator also provides a prelude to the discussion of coupled nonlinear systems. Suppose that the quadratic oscillator potential yields a stable frequency ω_0. If the nonlinear quartic term is small, it is natural to seek a perturbative solution to the equation of motion, expanding in the strength of the quartic term. Unfortunately, the first-order correction has a term that not only oscillates but also increases *linearly* with t. Such behavior indicates that a resonant driving term leads to a *secular growth* in the coordinate. This conclusion violates a theorem that the dynamical motion for this problem remains bounded for all t. It indicates that the straightforward perturbative analysis fails when $\omega_0 t \approx 1$. A more powerful improved analysis includes a simultaneous shift in the frequency $\omega_0 \to \omega$, which eliminates the secular term and allows the perturbative approach to hold for much longer times. This calculation illustrates the importance of resonance in driving the perturbations of nonlinear systems. We present some numerical results for this interesting system.

A different and particularly useful prototype for periodic nonlinear motion is the planar pendulum that is familiar from freshman physics. We initially use the usual dynamical variables $(p, q) = (p_\theta, \theta)$, with θ the angle measured from the down position and p_θ the angular momentum (Fig. 1.1). The full equations of motion are highly nonlinear. This simple physical system is remarkably rich, for it has both unstable and stable equilibrium points (up and down) and both libration (oscillation) and rotation (over the top) types of motion. It serves to introduce the concept of a fixed point corresponding to stationary coordinates. In addition, the linearized stability analysis about the fixed points leads to the notion of a separatrix formed by an orbit through an unstable fixed point. In this example, the solutions have a qualitatively different character on the two sides of the separatrix. In the elementary analysis, of course, the equations are linearized about the stable fixed point where the pendulum hangs down, and one finds simple harmonic motion. With an additional damping term, the system will decay back to this stable fixed point; in this case the fixed point is known as an attractor.

The action-angle variables (J, ϕ) will be seen to simplify and unify the dynamics of nonlinear periodic hamiltonian systems. We start by studying such a system with one degree of freedom. For both rotational and librational motion, the action J is a *constant*, and the angle variable increases *linearly* with the time. No matter how complicated the dynamics, the dynamical trajectory in this two-dimensional (J, ϕ) phase space is simply uniform motion along a straight line. As particular and important examples, we develop the action-angle description of the simple harmonic oscillator and the pendulum.

4

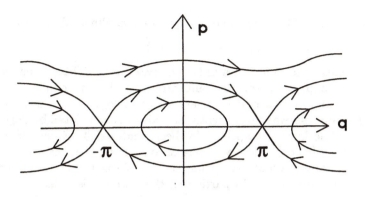

Fig. 1.1. Sketch of phase-space orbits for a simple pendulum with $(p, q) = (p_\theta, \theta)$. The closed orbits describe stable oscillations (librations); for small displacements, energy conservation implies that they form ellipses. The top orbit is a rotation where the pendulum goes over the top and the angle increases continuously. The origin is a stable fixed point, and the crossing points on the q axis at $\pm\pi$ are unstable fixed points where the pendulum points up; an orbit through them is known as a separatrix. Since θ is an angle, the figure is periodic in q with period 2π.

Figure 1.1 is a sketch of the phase-space orbits for the simple pendulum with $(p, q) = (p_\theta, \theta)$. It illustrates a typical phase space for a hamiltonian system, where the number of coordinates is necessarily even. The concept of phase space is more general, however, and applies to any system of coupled first-order differential equations. As an important and interesting example, we consider the Lorenz equations. They are a discrete first-order dynamical system with three dependent variables and a real non-negative parameter r that represents the rate of heating in the Rayleigh-Bénard problem. We first find the fixed points of the Lorenz equations, whose location depends explicitly on r. The linearized solutions around each fixed point serve to characterize the stability for the associated small-amplitude motion. As the parameter r increases, the solutions of the Lorenz equations progress from (1) static thermal conduction to (2) steady convective flow, followed by (3) periodic oscillations, and then (4) chaotic motion. Remarkably, far into the chaotic regime, one finds intervals of periodic motion and period doubling where the frequency decreases discontinuously by powers of two. We examine various numerical solutions, which readers can easily reproduce and extend for themselves. Two theorems are proven concerning the phase-space convergence of the solutions to the Lorenz equations [Sp82]: (1) a phase orbit eventually enters a bounded ellipsoid, and (2) the phase-space volume shrinks along a phase trajectory.

Finite-difference equations appear frequently as approximations to differential equations, although they are identical only in the limit of vanishing finite-difference spacing. Great insight into the general behavior of the solutions to the

Lorenz equations is obtained from one very simple nonlinear difference equation

$$x_{n+1} = \rho x_n (1 - x_n) \tag{1.1}$$

where ρ is a parameter and x lies in the interval 0 to 1. This system is known as the logistic map. In 1976, May summarized its behavior in a widely read article [Ma76] and emphasized its relevance to biology, economics and social sciences. The analysis was subsequently extended in some detail by Feigenbaum [Fe78, Fe80]. The logistic map has also been analyzed by Kadanoff in a very accessible article [Ka83]. Here we present some numerical results, as well as simple analytical results that clearly illustrate the stable fixed point, bifurcation, period doubling, and chaos. In particular, the characteristic features of chaos are:

- apparently random behavior

- actual determinism

- extreme sensitivity to initial conditions

The many-body problem of the planets in their keplerian orbits around the sun involves coupled nonlinear systems, each one of which would be separable and periodic if there were no other planets [Ar89a]. As a basis for subsequent discussion, we review the Hamilton-Jacobi theory of separable periodic systems, relying on the action-angle variables (J, ϕ). These particular canonical variables provide a very useful basis to characterize the dynamics of such coupled hamiltonian systems. Our previous action-angle description of a single periodic system (for example, the planar pendulum) is then extended to separable, integrable periodic systems with two degrees of freedom. Although the full dynamical phase space is four dimensional, the conservation of energy for each system restricts the dynamical motion to a two-dimensional toroidal phase space.[2]

We then turn to the problem of the *coupled* motion of separable periodic systems. As a very simple introductory model [Ru86, Pe99], we consider the periodic motion of a single system with an additional time-dependent interaction

$$\varepsilon V(J, \phi, t) \tag{1.2}$$

where ε is a small dimensionless parameter and V is time periodic with period 2π in the (rescaled) time. Such a perturbation deforms the originally straight trajectories in the (J, ϕ) plane by corrections of order ε. It is remarkable that an *additional* canonical transformation containing a term of order ε to new action-angle variables $(\tilde{J}, \tilde{\phi})$ can re-straighten the action-angle orbits throughout most of phase space. This procedure will fail, however, near one or more of a denumerably infinite set of points determined by the resonance condition

$$m\omega(J) - n = 0 \qquad ; \; m, n \text{ integers} \tag{1.3}$$

---

[2]Along the way we provide a proof of Liouville's theorem, which states that for a hamiltonian system the multidimensional phase-space volume is preserved along a phase trajectory.

where $\omega(J)$ is the uncoupled angular frequency. For a given pair of integers m and n, this equation defines a resonant value J_r of the action.

A direct analysis of the new hamiltonian in the vicinity of a resonance leads to a pendulum equation in the new action-angle variables with all the associated behavior described above in Fig. 1.1. We shall observe that "near" means to distances $\sim \sqrt{\varepsilon}$. Thus in the phase space of the new action-angle variables $(\tilde{J}, \tilde{\phi})$, one has "undisrupted" regions where the trajectories are straight lines to order ε, and "disrupted" regions centered on the resonant values of the action J_r, where the plots look like those in Fig. 1.1. In the latter regions, the phase space contains islands of stability and separatrix crossings on the axis. For a given J_r, the size of the disrupted region is $|\tilde{J} - J_r| \sim \sqrt{\varepsilon}$.

A conceptually similar problem is two weakly coupled degrees of freedom, each undergoing periodic motion [Pe99]. For given total energy, and given value of some other combination of (J_1, J_2), the periodic coordinates (ϕ_1, ϕ_2) in the uncoupled problem lie in a closed surface (a "two-torus") and are linear functions of the time. If the ratio of uncoupled frequencies $\omega_1(J_1)/\omega_2(J_2)$ is incommensurate, the phase trajectory eventually covers the whole torus, coming arbitrarily close to any given point; in contrast, if this ratio is rational, the trajectory eventually closes on itself. For different initial conditions, the two actions and the corresponding two frequencies change, leading to a set of nested tori determined by the actions (J_1, J_2).

Now turn on a coupling term

$$\varepsilon V(J_1, J_2, \phi_1, \phi_2) \tag{1.4}$$

As the actions change, the corresponding frequencies change. The condition $\omega_1/\omega_2 = n/m$ that gives commensurate frequencies again will lead to a resonant disruption of the phase space, with disrupted tori sandwiched between undisrupted ones. This behavior is readily seen computationally, where multidimensional phase trajectories are studied with Poincaré sections constructed by recording those points where the phase trajectory passed through a particular plane.

How large is the total disrupted region? Is it finite, or does it fill all of phase space? The answer is given by the KAM theorem. We do not prove this theorem, but we are now at least in a position to understand it. The KAM theorem states that, for sufficiently small ε, the sum of all disrupted regions is a small fraction of the total phase space that vanishes as $\varepsilon \to 0$. For a careful description of this celebrated theorem and the subtleties associated with its proof, see, for example, [Ar89a, Li92, Sc94, Jo98a, Ot02a].

For two degrees of freedom with constant energy, the phase-space motion remains confined by adjacent undisrupted tori until ε grows sufficiently to destroy the structure of phase space. For three or more degrees of freedom, the situation becomes more complicated. One can find chaotic trajectories that extend throughout phase space as well as islands of stability. This behavior is observed computationally through the Poincaré sections.

We very much hope that this supplement on nonlinear dynamics will enable readers to approach the literature on contemporary mechanics and carry out their own computational explorations of this fascinating new subject that lies at the interface between physics and mathematics.

Part II
Nonlinear Continuous Systems

2 Linearized stability analysis

Chapter 10 in [Fe03] is titled *Surface waves on fluids*. If the fluid is incompressible and the motion irrotational, then the basic equation of hydrodynamics with a pressure force and gravity can be integrated over all space to yield Bernoulli's Eq. (54.6) in the interior of the fluid[3]

$$-\frac{\partial \Phi}{\partial t} + \frac{1}{2}v^2 + U + \frac{p}{\rho} = 0 \tag{2.1}$$

Here $\mathbf{v} = -\boldsymbol{\nabla}\Phi$ is the velocity field, p is the pressure, ρ the mass density, and $U = gz$ is the gravitational potential. Since the fluid is incompressible, the continuity equation $\partial\rho/\partial t + \boldsymbol{\nabla} \cdot (\rho\mathbf{v}) = 0$ implies that the velocity potential satisfies Laplace's equation everywhere in the interior

$$\nabla^2 \Phi = 0 \tag{2.2}$$

On any fixed boundary, the normal component of the fluid velocity must vanish

$$\hat{\mathbf{n}} \cdot \boldsymbol{\nabla}\Phi = 0 \qquad \text{; fixed boundary} \tag{2.3}$$

The hydrodynamic equations are intrinsically nonlinear. To motivate the subsequent material, we examine the stability of two familiar physical systems by linearizing the relevant equations about a steady equilibrium configuration. As in previous analyses, we consider the normal-mode solutions to the linearized equations.

Rayleigh-Taylor instability

Suppose that one has two fluids, one on top with properties labeled by (1) and the other on the bottom with properties labeled by (2), in a closed container (Fig. 2.1). This is the configuration considered in Problem 10.8 in [Fe03]; for a generalization to viscous fluids, see [Ch81a].

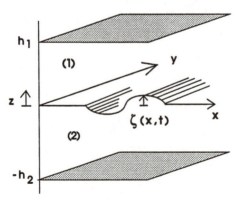

Fig. 2.1. Sketch of physical configuration in Problem 10.8.

---

[3]Equations and figures with a prefix greater than 15 refer to [Fe03].

Let the displacement of the interface in the vertical direction be labeled by ζ. If the fluid does not move, then the pressure just arises from the piled up fluid, and integration of Newton's law in the z direction provides the static pressure for this configuration [these equations are direct generalizations of Eq. (53.5) for the present case of two superposed fluids]

$$
\begin{aligned}
p_1^0 &= \rho_1 g (h_1 - z) & &; \zeta \le z \le h_1 \\
p_2^0 &= \rho_1 g (h_1 - \zeta) + \rho_2 g (\zeta - z) & &; -h_2 \le z \le \zeta \qquad (2.4)
\end{aligned}
$$

The pressure is continuous across the interface; the corresponding force, which arises from the gradient of the pressure, is not. Of course, given such a configuration, the fluid will move and there will be a velocity field. Assume that the surface displacement is small, and work to first order in the displacement ζ and velocity $\mathbf{v}$. There will be a first-order induced pressure δp in the fluids that drives the velocity field, and Bernoulli's equations in the interior of the two fluids read

$$
\begin{aligned}
-\rho_1 \frac{\partial \Phi_1}{\partial t} &= -p_1^0 - \delta p_1 - \rho_1 U \\
-\rho_2 \frac{\partial \Phi_2}{\partial t} &= -p_2^0 - \delta p_2 - \rho_2 U \qquad (2.5)
\end{aligned}
$$

Now take the difference of these two equations across the interface where $U = g\zeta$, and use the continuity of the pressure there

$$
\left[-\rho_1 \frac{\partial \Phi_1}{\partial t} + \rho_2 \frac{\partial \Phi_2}{\partial t} \right]_{z=\zeta} = g(\rho_2 - \rho_1)\zeta \qquad (2.6)
$$

Since Φ is already of first order, the left hand side can be evaluated at $z = 0$.[4]

To first order, the vertical velocity $\partial\zeta/\partial t$ of the interface is just the vertical component of the fluid velocity, and one has the boundary condition at the interface [Eq. (54.20)]

$$
v_z = -\frac{\partial \Phi_1}{\partial z} = -\frac{\partial \Phi_2}{\partial z} = \frac{\partial \zeta}{\partial t} \qquad (2.7)
$$

evaluated at $z = 0$. Hence the partial time derivative of Eq. (2.6) yields the dynamical equation for the interface re-expressed in terms of the velocity potentials

$$
\rho_2 \frac{\partial^2 \Phi_2}{\partial t^2} - \rho_1 \frac{\partial^2 \Phi_1}{\partial t^2} = -g(\rho_2 - \rho_1)\frac{\partial \Phi}{\partial z} \qquad (2.8)
$$

where all quantities are now evaluated at $z = 0$ (the preceding equation ensures that either Φ_1 or Φ_2 can be used on the right hand side). Note that the pressure

---

[4]Note that any constant term in Bernoulli's equation, appearing, for example, from the choice of zero in the potential or pressure, can be removed with a gauge transformation $\Phi \to \Phi + \text{constant} \times t$. We assume that this has been done so that now Φ is explicitly of first order.

change δp has disappeared from this relation; it does not have to be known beforehand to find the solution! In this way, the problem reduces to finding solutions to Laplace's equation in the upper and lower fluids that satisfy the fixed surface boundary conditions, along with Eq. (2.8) at the interface. The surface profile then follows from Eq. (2.6).

We simplify the physical configuration and look for simple harmonic surface waves that are independent of y and propagate in the x direction (Fig. 2.1); a general solution can be obtained by superposing such waves. The reader can easily confirm that the following velocity potentials satisfy the conditions Eqs. (2.2) and (2.3)

$$\Phi_1(x, z, t) = \Phi_0 \frac{\cosh[k(z - h_1)]}{\sinh[k(-h_1)]} \exp[i(kx - \omega t)]$$

$$\Phi_2(x, z, t) = \Phi_0 \frac{\cosh[k(z + h_2)]}{\sinh[k(h_2)]} \exp[i(kx - \omega t)] \tag{2.9}$$

where the real part is taken at the end of the analysis. The quantity Φ_0 is some small, first-order (complex) amplitude and the coefficients are chosen to yield the same vertical velocity at the interface, as indicated in Eq. (2.7). Substituting these amplitudes into the dynamical equation (2.8) determines the dispersion relation $\omega(k)$ that fixes the allowed frequency for a given wavenumber k, which we assume to be positive. The time dependence cancels upon substitution (these are *normal modes*), and the result is[5]

$$\omega^2(k) = \frac{gk(\rho_2 - \rho_1)}{\rho_2 \coth(kh_2) + \rho_1 \coth(kh_1)} \tag{2.10}$$

This is the answer given in Prob. 10.8.

How are these results to be interpreted? If the bottom fluid is heavier, so that $\rho_2 > \rho_1$, then Eq. (2.10) shows that $\omega^2(k)$ is positive for all k. This implies that the small-amplitude disturbance is a propagating surface wave, and that the static situation is stable against small perturbations. Note the two interesting limiting cases

1. If $kh_1 \gg 1$ and $kh_2 \gg 1$ (short wavelengths), then the hyperbolic cotangents approach 1, and the dispersion relation becomes

$$\omega^2(k) \approx \frac{gk\,(\rho_2 - \rho_1)}{\rho_2 + \rho_1} \tag{2.11}$$

This result has the same form as that found in Eq. (54.34) for waves on a deep fluid (but now involving the relative fluid density)

2. If $kh_1 \ll 1$ and $kh_2 \ll 1$ (long wavelengths), then the hyperbolic cotangents approach the inverse arguments, and the dispersion relation becomes

$$\omega^2(k) \approx \frac{gk^2\,(\rho_2 - \rho_1)}{\rho_2/h_2 + \rho_1/h_1} \tag{2.12}$$

[5]Note that $\coth(-x) = -\coth(x)$ is odd in its argument. This behavior ensures that $\omega^2(k)$ in Eq. (2.10) is an even function of k. For the subsequent limiting cases such as Eq. (2.11), k should strictly be interpreted as $|k|$.

This result has the same form as that found in Eq. (54.33) for waves on a shallow fluid.

To understand both the general behavior and these particular limits, note that a solution of Laplace's equation cannot have spatial oscillations in all directions, because the sum of the curvatures must vanish. In particular, Eq. (2.9) shows that the wave amplitude oscillates in the x direction and decays exponentially away from the interface like $\sim \exp(-|kz|)$. Consequently, the oscillatory motion is confined to a surface region of thickness $\sim k^{-1} = \lambda/2\pi$, where λ is the wavelength. In the short-wavelength limit, the confining planes are many wavelengths away and play no role, so that Eq. (2.11) is independent of h_i; for long wavelengths, in contrast, the motion is essentially independent of z, and the boundaries predominate, as seen in Eq. (2.12).

Suppose, on the other hand, that the heavier fluid is on top so that $\rho_2 < \rho_1$. Then Eq. (2.10) implies that $\omega^2(k) < 0$ for all k. In this case, we define

$$\omega^2(k) \equiv -\gamma_k^2 \qquad\qquad ; \gamma_k^2 > 0 \qquad\qquad (2.13)$$

If Eq. (2.13) holds, then these solutions behave like

$$\exp(ikx)\exp(\pm\gamma_k t) \qquad ; \gamma_k > 0 \qquad\qquad (2.14)$$

In particular, the solution proportional to $\exp(+\gamma_k t)$ *grows exponentially with time* and soon violates the condition of a small, first-order disturbance. This solution illustrates the *Rayleigh-Taylor gravitational instability*. When the heavier fluid is on top, the initial static solution is unstable with respect to modes with any wave number k. The corresponding growth rate is γ_k, which increases with increasing k; thus short-wavelength disturbances grow faster than those with long wavelengths.

The effect of surface tension can be readily included in these results. If τ is the surface tension of the interface (see Fig. 54.6), then there is an additional pressure difference across the interface given by Eq. (54.77)[6]

$$p_2 - p_1 = -\tau\nabla^2\zeta \qquad\qquad (2.15)$$

Hence there will now be an additional term in the difference in Eq. (2.6)

$$\left[-\rho_1\frac{\partial\Phi_1}{\partial t} + \rho_2\frac{\partial\Phi_2}{\partial t}\right]_{z=0} = g(\rho_2 - \rho_1)\zeta - \tau\nabla^2\zeta \qquad\qquad (2.16)$$

For the configuration in Fig. 2.1, the last term becomes $-\tau\,\partial^2\zeta/\partial x^2$. A partial time derivative then leads to the appropriate extension of Eq. (2.8)

$$\rho_2\frac{\partial^2\Phi_2}{\partial t^2} - \rho_1\frac{\partial^2\Phi_1}{\partial t^2} = -\left[g(\rho_2 - \rho_1) - \tau\frac{\partial^2}{\partial x^2}\right]\frac{\partial\Phi}{\partial z} \qquad\qquad (2.17)$$

---

[6]Note this term can actually be written as $-\tau\nabla_\perp^2\zeta = -\tau(\partial^2/\partial x^2 + \partial^2/\partial y^2)\zeta(x,y,t)$ since ζ only depends on these variables.

Substitution of the solution in Eq. (2.9) yields the generalization of Eq. (2.10)

$$\omega^2(k) = \frac{gk(\rho_2 - \rho_1) + \tau k^3}{\rho_2 \coth(kh_2) + \rho_1 \coth(kh_1)} \tag{2.18}$$

For the simplified case of a vacuum-fluid interface $[\rho_1 = 0$ and $(\rho_2, h_2) \equiv (\rho, h)]$, one obtains the dispersion relation given in Prob. 10.9

$$\omega^2(k) = \frac{gk\rho + \tau k^3}{\rho \coth(kh)} = \left(gk + \frac{\tau k^3}{\rho}\right) \tanh(kh) \tag{2.19}$$

When the heavier fluid is below the lighter fluid ($\rho_2 > \rho_1$), Eq. (2.18) shows that surface tension simply enhances the stability, and $\omega^2(k)$ remains positive for all k. When the heavier fluid is on top ($\rho_2 < \rho_1$), however, the situation becomes more interesting. In this case, the numerator of Eq. (2.18) indicates that surface tension stabilizes waves with sufficiently large k (namely, sufficiently small wavelengths[7]), for $\omega^2(k)$ becomes positive above a critical value of k given by

$$k_0^2 = \frac{g(\rho_1 - \rho_2)}{\tau} \tag{2.20}$$

For $k^2 < k_0^2$, the growth rate is determined by

$$\gamma_k^2 = \frac{g(\rho_1 - \rho_2)k - \tau k^3}{\rho_2 \coth(kh_2) + \rho_1 \coth(kh_1)} \tag{2.21}$$

Assume, to simplify the algebra, that the bounding surfaces are far away with $(h_1, h_2) \to \infty$, so that the denominator can be replaced by $\rho_2 + \rho_1$. The resulting γ_k^2 vanishes at k_0 and at $k = 0$. In between, γ_k^2 clearly has a maximum at k_c given by

$$k_c^2 = \frac{g(\rho_1 - \rho_2)}{3\tau} = \frac{1}{3}k_0^2 \tag{2.22}$$

with a maximum value of

$$\gamma_c^2 = \frac{2\tau k_c^3}{\rho_2 + \rho_1} \tag{2.23}$$

As expected from the previous example with no surface tension, k_c^2 and γ_c^2 both become large as $\tau \to 0$.

A classical system in thermal equilibrium has all normal modes excited with random amplitudes and phases. In this case, one expects that modes with wave number k_c (or nearby values) will dominate the time dependence because they have the largest growth rate γ_c. Hence the initial distortions of the surface will increase like

$$\exp\left(ik_c x\right) \exp\left(+\gamma_c t\right) \tag{2.24}$$

---

[7] Recall $k = 2\pi/\lambda$.

14

Note that this remarkably simple analysis provides both a characteristic *time scale* γ_c^{-1} for the growth of the instability and a characteristic *wave number* k_c (so that the initial surface distortion should have a ripple pattern with wavelength $\lambda_c = 2\pi/k_c$).

Kelvin-Helmholtz instability

We now extend the previous analysis to the case where the two fluids each move with *constant uniform horizontal velocities* $\mathbf{u}_1$ and $\mathbf{u}_2$ respectively.[8] The new feature here is that the velocity potential contains a zeroth-order part arising from the uniform flow, as well as the perturbation from the small-amplitude distortion. For simplicity, it is convenient to assume that both $\mathbf{u}_1$ and $\mathbf{u}_2$ are in the x direction.[9] In addition, we assume that each fluid is semi-infinite, so that $(h_1, h_2) \to \infty$.

Since the fluid is incompressible, the velocity potential continues to satisfy Laplace's equation (2.2). For a wave propagating along the x direction, the relevant solutions are

$$\begin{aligned}
\Phi_1(x, z, t) &= -u_1 x + \Phi_{10}\, e^{-kz} \exp\left[i\left(kx - \omega t\right)\right] \\
\Phi_2(x, z, t) &= -u_2 x + \Phi_{20}\, e^{kz} \exp\left[i\left(kx - \omega t\right)\right]
\end{aligned} \tag{2.25}$$

Here Φ_{10} and Φ_{20} are small first-order amplitudes, and the real part is implied throughout. The resulting velocity components are

$$\begin{aligned}
v_{x1} &= -\frac{\partial \Phi_1}{\partial x} = u_1 - ik\,\Phi_{10}\, e^{-kz} \exp\left[i\left(kx - \omega t\right)\right] \\
v_{z1} &= -\frac{\partial \Phi_1}{\partial z} = k\,\Phi_{10}\, e^{-kz} \exp\left[i\left(kx - \omega t\right)\right]
\end{aligned} \tag{2.26}$$

with similar expressions for the second fluid. In addition, the surface profile itself is a traveling wave with

$$\zeta(x, t) = \zeta_0 \exp\left[i\left(kx - \omega t\right)\right] \tag{2.27}$$

Bernoulli's equation (2.1) in medium 1 involves $\mathbf{v}_1^2/2$, which leads to the linearized quantity

$$\frac{1}{2}\mathbf{v}_1^2 \approx \frac{1}{2}u_1^2 - iku_1\Phi_{10}e^{-kz}\exp[i(kx - \omega t)] \tag{2.28}$$

Here, the constant zeroth-order contribution $u_1^2/2$ can be eliminated with a gauge transformation.

To incorporate the uniform flow, it is convenient to retain explicitly the three first-order amplitudes Φ_{10}, Φ_{20} and ζ_0, instead of eliminating ζ_0 as in

---

[8] A horizontal velocity has $\mathbf{u} = (u_x, u_y, 0)$.
[9] A treatment with arbitrary $\mathbf{u}_1$ and $\mathbf{u}_2$ shows that the criterion for the onset of instability in Eq. (2.42) merely involves the more general vector difference $|\mathbf{u}_2 - \mathbf{u}_1|^4$.

Eq. (2.8). The resulting linearized Bernoulli's equation for fluid 1, evaluated at the interface, yields

$$-\rho_1 \left(\frac{\partial}{\partial t} + u_1 \frac{\partial}{\partial x} \right) \Phi_1(x, z = 0, t) = -\delta p_1(x, z = 0, t) - \rho_1 g \zeta(x, t) \quad (2.29)$$

where $\delta p_1(x, z = 0, t)$ now contains all first-order corrections.[10] In the present case of a traveling surface wave, we find

$$
\begin{aligned}
\delta p_1 &= \rho_1 \left[-i \left(\omega - k u_1 \right) \Phi_{10} - g \zeta_0 \right] \exp \left[i \left(kx - \omega t \right) \right] \\
&= \delta p_{10} \exp \left[i \left(kx - \omega t \right) \right]
\end{aligned}
\quad (2.30)
$$

where

$$\delta p_{10} = -\rho_1 \left[i \left(\omega - k u_1 \right) \Phi_{10} + g \zeta_0 \right] \quad (2.31)$$

is a small first-order amplitude.

Bernoulli's equation for the second fluid yields a similar pressure

$$
\begin{aligned}
\delta p_2 &= \delta p_{20} \exp \left[i \left(kx - \omega t \right) \right] \\
\delta p_{20} &= -\rho_2 \left[i \left(\omega - k u_2 \right) \Phi_{20} + g \zeta_0 \right]
\end{aligned}
\quad (2.32)
$$

Surface tension determines the difference in the pressure $\delta p_2 - \delta p_1$ across the interface according to Eq. (2.15), which here yields

$$-i \rho_2 \left(\omega - k u_2 \right) \Phi_{20} - \rho_2 g \zeta_0 + i \rho_1 \left(\omega - k u_1 \right) \Phi_{10} + \rho_1 g \zeta_0 = \tau k^2 \zeta_0 \quad (2.33)$$

This equation provides one relation between the three amplitudes Φ_{20}, Φ_{10} and ζ_0.

The boundary condition at the interface must now be revised and extended. It is shown in Eq. (54.19) in [Fe03] that the correct boundary condition for the motion of the fluid at the surface is

$$v_z = -\frac{\partial \Phi}{\partial z} = \frac{\partial \zeta}{\partial t} + \mathbf{v} \cdot \boldsymbol{\nabla} \zeta \quad (2.34)$$

The right hand side contains the hydrodynamic (or substantive) derivative of the displacement, where $\mathbf{v}$ is the total velocity. In the present case of a wave propagating along x, Eq. (2.26) shows that the last term becomes $u_j \partial \zeta / \partial x$ to first order, where $j = 1$ or 2 for the upper or lower fluid. Hence the surface boundary condition is now different on the two sides of the interface

$$
\begin{aligned}
v_{z1} &= -\frac{\partial \Phi_1}{\partial z} = \frac{\partial \zeta}{\partial t} + u_1 \frac{\partial \zeta}{\partial x} \\
v_{z2} &= -\frac{\partial \Phi_2}{\partial z} = \frac{\partial \zeta}{\partial t} + u_2 \frac{\partial \zeta}{\partial x}
\end{aligned}
\quad (2.35)
$$

[10]Note that this approach differs from that in Eqs. (2.5); it clearly has no effect on the final difference in Eq. (2.33).

Use of Eqs. (2.25) and (2.27) yields the following two relations for the coupled amplitudes

$$
\begin{aligned}
k\Phi_{10} &= -i\left(\omega - ku_1\right)\zeta_0 \\
k\Phi_{20} &= i\left(\omega - ku_2\right)\zeta_0
\end{aligned}
\tag{2.36}
$$

A combination with Eq. (2.33) shows that the angular frequency must satisfy the dispersion relation

$$
\rho_2\left(\omega - ku_2\right)^2 + \rho_1\left(\omega - ku_1\right)^2 = g(\rho_2 - \rho_1)k + \tau k^3
\tag{2.37}
$$

This relation is the basis for Prob. 10.10. It evidently reduces to Eq. (2.18) when $u_1 = u_2 = 0$ [after taking the limits $(h_1, h_2) \to \infty$].

How do we interpret these results? Consider the case where $\rho_2 > \rho_1$ so that the *heavier fluid is on the bottom*. This configuration is gravitationally stable if the fluids are initially at rest (namely if $u_1 = 0$ and $u_2 = 0$). Here, the dispersion relation in Eq. (2.37) provides a quadratic equation for the angular frequency

$$
\omega^2 - \frac{2k(\rho_1 u_1 + \rho_2 u_2)}{\rho_1 + \rho_2}\omega + \frac{k^2(\rho_1 u_1^2 + \rho_2 u_2^2)}{\rho_1 + \rho_2} - \frac{gk(\rho_2 - \rho_1)}{\rho_1 + \rho_2} - \frac{\tau k^3}{\rho_1 + \rho_2} = 0
\tag{2.38}
$$

With a little algebra, the solution to this quadratic equation is obtained as

$$
\omega(k) = \frac{k(\rho_1 u_1 + \rho_2 u_2)}{\rho_1 + \rho_2} \pm \left[\frac{gk(\rho_2 - \rho_1)}{\rho_1 + \rho_2} + \frac{\tau k^3}{\rho_1 + \rho_2} - \frac{\rho_1 \rho_2 (u_1 - u_2)^2 k^2}{(\rho_1 + \rho_2)^2}\right]^{1/2}
\tag{2.39}
$$

The configuration will be stable if ω is real for all k, for then the small-amplitude normal modes will simply be propagating surface waves. The condition for real ω is that the expression inside the square root be positive

$$
\frac{\rho_1 \rho_2 (u_1 - u_2)^2}{\rho_1 + \rho_2} < \frac{g(\rho_2 - \rho_1)}{k} + k\tau
\tag{2.40}
$$

This inequality is always satisfied at small enough or large enough k (large enough or small enough wavelength). The minimum value of the right hand side is evidently reached at

$$
k_c^2 = \frac{g(\rho_2 - \rho_1)}{\tau}
\tag{2.41}
$$

Insertion of this value on the right hand side in Eq. (2.40) yields the inequality

$$
(u_1 - u_2)^4 < \frac{4\tau g(\rho_2 - \rho_1)(\rho_1 + \rho_2)^2}{\rho_1^2 \rho_2^2}
\tag{2.42}
$$

If the relative velocity of the fluids satisfies this inequality (so that the relative velocity is not too large), then there is no value of k that violates the inequality in Eq. (2.40). Hence the *configuration is stable* for all k, which means for all normal-mode excitations. This is the answer given in Prob. 10.10.

If the inequality in Eq. (2.42) is violated, then $\omega(k)$ becomes complex, and the configuration is unstable with respect to some normal-mode excitations. With increasing $(u_1 - u_2)^2$, the onset of this *Kelvin-Helmholtz instability* will first occur for $k = k_c$, with vanishingly small growth rate. If the inequality is slightly violated, one gets a band of unstable modes in the vicinity of k_c, and the central one will grow most rapidly. Once again this analysis yields both the time scale for the growth rate of this instability and its spatial pattern. There are several interesting limiting cases

1. If there is no surface tension ($\tau = 0$), then the system is unstable for arbitrarily small relative transverse velocities.

2. If the upper fluid is near vacuum ($\rho_1 \approx 0$) with positive τ, then the bottom fluid is always stable no matter how large u_2 becomes.

3. This analysis applies to moving air and stationary water, where $\rho_1 \approx 1.3 \times 10^{-3}$ g cm^{-3} and $\rho_2 \approx 1.0$ g cm^{-3}. The measured value $\tau \approx 75$ dyne/cm and $g \approx 980$ cm s^{-2} yield the critical air speed $u_{1c} \approx 650$ cm/s, which is 23 km/hr or 15 mi/hr. It is known that the onset of white caps on the ocean occurs near this wind speed [Ch81b].

For a stationary configuration, Bernoulli's Eq. (2.1) implies that an increase in velocity at a fixed height implies a decrease in pressure (and *vice versa*). The physics of this Kelvin-Helmholtz instability is that an increase in the velocity difference between the fluids eventually produces a surface "lift" that neither gravity nor surface tension can compensate.

3   Rayleigh-Bénard problem: basic formulation

Chapter 11 in [Fe03] is on *Heat conduction*, as found typically in solids. As discussed in Chap. 12 there, heat transfer in a fluid can also occur through *convection*. In the absence of internal heat sources, and with the neglect of viscous heating, the temperature in isobaric incompressible fluid flow satisfies Eq. (60.48)

$$\frac{\partial T}{\partial t} + \mathbf{v} \cdot \boldsymbol{\nabla} T = \kappa \nabla^2 T \tag{3.1}$$

Here the thermal diffusivity κ is defined in Eq. (57.13) as $\kappa = k_{th}/\rho c_p$ where k_{th} is the thermal conductivity, and c_p is the constant-pressure heat capacity per unit mass.

Chapter 12 in [Fe03] is on *Viscous fluids*, where the Navier-Stokes equation is derived. It states that for incompressible flow in a gravitational field $\mathbf{f} = -g\hat{\mathbf{z}}$, the fluid velocity satisfies Eq. (60.27)

$$\frac{d\mathbf{v}}{dt} = \frac{\partial \mathbf{v}}{\partial t} + (\mathbf{v} \cdot \boldsymbol{\nabla})\mathbf{v} = -\frac{1}{\rho}\boldsymbol{\nabla}p - g\hat{\mathbf{z}} + \nu\nabla^2\mathbf{v} \tag{3.2}$$

where ν is the kinematic viscosity (see Table 60.1).

We shall here concentrate on the particular problem of the onset of convection in a fluid heated from below. In this case, thermal expansion of the hotter fluid at the bottom creates an unstable density gradient (the Rayleigh-Taylor instability discussed in Sec. 2 above for the simpler case of a nonviscous fluid). For a viscous fluid, however, the onset of the convective instability arises from a balance between the buoyant force on the lower less dense fluid and the viscous force that acts to maintain the static equilibrium. To include this physical effect, the driving term from the pressure $-(1/\rho)\boldsymbol{\nabla}p$ in Eq. (3.2) must incorporate the density change caused by the heating of the fluid. Otherwise, we continue to assume incompressible flow, so that

$$\boldsymbol{\nabla} \cdot \mathbf{v} = 0 \tag{3.3}$$

We now have five coupled, nonlinear differential equations for the five functions $(\mathbf{v}, p, T)$.[11] Various solutions to these equations under simplifying assumptions are examined in [Fe03]. In this section, we investigate some additional consequences of these general equations.

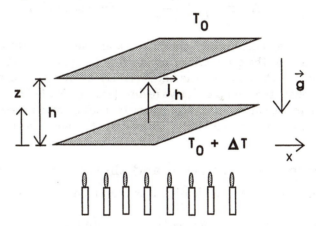

Fig. 3.1. Sketch of the physical configuration considered in this section: a fluid between two fixed horizontal surfaces a distance h apart, heated from below.

We focus here on the physics of a particular configuration, that of a fluid confined between two parallel horizontal surfaces separated by a distance h and with the lower surface maintained at a higher temperature $T_0 + \Delta T$ than the upper one at T_0 (Fig. 3.1). Initially, at small enough h and ΔT, the heat flow between the plates will be purely conductive. In this case, the fluid is at rest, and

---

[11] These partial differential equations are first-order in time and second-order in space.

the velocity field $\mathbf{v}$ will vanish. For this static conductive state, the equilibrium temperature $T^0(z)$ obeys Laplace's equation [see Eq. (3.1)] and varies linearly with the distance between the plates

$$T^0(z) = T_0 + \frac{\Delta T}{h}(h - z) \tag{3.4}$$

The general expression for the conductive heat flux $\mathbf{j}_h$ is given by Eq. (57.11)

$$\mathbf{j}_h = -k_{\text{th}} \boldsymbol{\nabla} T \tag{3.5}$$

For the static configuration in Fig. 3.1 and Eq. (3.4), this quantity will be given by

$$\mathbf{j}_h = k_{\text{th}} \frac{\Delta T}{h} \hat{\mathbf{z}} \tag{3.6}$$

Suppose now that ΔT is increased. At a critical value R_c of the dimensionless Rayleigh number $R = gh^3 \beta \Delta T / \nu\kappa$ [see Eq. (3.23) below], convection will set in. This extra convective mass and energy transfer *increases* the total heat flow relative to that in the pure conductive state, and the excess heat transfer will grow from zero with increasing $R - R_c > 0$. One can define the dimensionless Nusselt number

$$N \equiv \frac{(\mathbf{j}_h^{\text{tot}})_z}{k_{\text{th}} \Delta T / h} \quad ; \text{ Nusselt number} \tag{3.7}$$

where $(\mathbf{j}_h^{\text{tot}})_z$ now includes both conductive and convective heat transfer. If N is plotted against R, the conductive state for $R \leq R_c$ appears as the horizontal line $N = 1$. For $R > R_c$, however, the line $N(R)$ has a finite slope characterizing the combined conductive and convective regime (see Fig. 5.1 below). This change in slope provides a clear experimental means of distinguishing these distinct states and the transition between them. Ultimately, the flow becomes turbulent for $R \gg R_c$.

Boussinesq approximation and thermal expansion

For small changes in the density, it is convenient to write

$$\rho = \rho_0 \left[1 - \beta(T - T_0) \right] \tag{3.8}$$

where $\rho_0 = \rho(T_0)$ is the density at some reference temperature T_0 (chosen as that of the upper plate). Here, β is the thermal expansion coefficient

$$\beta = -\frac{1}{\rho} \left(\frac{\partial \rho}{\partial T} \right)_p = \frac{1}{V} \left(\frac{\partial V}{\partial T} \right)_p \tag{3.9}$$

and the last equality follows from the definition of the mass density $\rho = M/V$. The physical constant β is typically positive and of order $10^{-3} \, (^\circ\text{C})^{-1}$ (for water at 20°C, experiments give $\beta \approx 2 \times 10^{-3} \, (^\circ\text{C})^{-1}$ [Ch81c]). The thermally induced change in the density is of order $\Delta\rho_{\text{th}} \sim \rho_0 \beta \Delta T$.

There is also a change in the density arising from the change in pressure, for example

$$\rho_{\text{press}} - \rho_0 = \left(\frac{\partial\rho}{\partial p}\right)_S (p - p_0) = \frac{p - p_0}{c_s^2} \tag{3.10}$$

where c_s is the speed of sound. Since the change in pressure from the bottom to the top in Fig. 3.1 is $|\Delta p| = \rho_0 g h$, one has

$$|\Delta\rho_{\text{press}}| \ll |\Delta\rho_{\text{th}}|$$

$$\text{provided} \quad \frac{gh}{c_s^2} \ll \beta\Delta T \tag{3.11}$$

This inequality is always satisfied for small enough h and large enough c_s^2, which we assume to hold.[12]

We now focus on the pressure driving term in the Navier-Stokes equation (3.2). For small temperature gradient, we use Eq. (3.8) to rewrite this contribution as

$$\frac{1}{\rho}\boldsymbol{\nabla}p = \frac{1}{\rho_0\left[1 - \beta\left(T - T_0\right)\right]}\boldsymbol{\nabla}p$$

$$\approx \frac{1}{\rho_0}\boldsymbol{\nabla}p + \frac{\beta\left(T - T_0\right)}{\rho_0}\boldsymbol{\nabla}p \tag{3.12}$$

To leading order, the pressure p in Eq. (3.2) simply reflects the piled-up fluid, so that we can set $\boldsymbol{\nabla}p/\rho_0 \approx -g\hat{\mathbf{z}}$ in the correction (second) term. In this way, Eq. (3.12) takes the form

$$\frac{1}{\rho}\boldsymbol{\nabla}p = \frac{1}{\rho_0}\boldsymbol{\nabla}p - \beta\left(T - T_0\right)g\hat{\mathbf{z}} \tag{3.13}$$

A combination with Eq. (3.2) yields a modified form of the Navier-Stokes equation that incorporates the buoyancy force arising from the thermal expansion

$$\frac{\partial\mathbf{v}}{\partial t} + (\mathbf{v}\cdot\boldsymbol{\nabla})\mathbf{v} = -\frac{1}{\rho_0}\boldsymbol{\nabla}p + g\beta\left(T - T_0\right)\hat{\mathbf{z}} - g\hat{\mathbf{z}} + \nu\nabla^2\mathbf{v} \tag{3.14}$$

Equations (3.1), (3.3), and (3.14) provide five coupled, nonlinear equations for the five quantities $(\mathbf{v}, p, T)$.[13] They constitute the *Boussinesq approximation* to the Rayleigh-Bénard problem. These equations depend on three separate and independent physical parameters of the fluid (ν, κ, β). The only assumptions are those involved in Eqs. (3.1)-(3.3), along with the linear dependence of ρ/ρ_0 on the small temperature difference in Eq. (3.8). In addition, the quantities

---

[12]Note that the compressibility $(1/\rho)(\partial\rho/\partial p)$ is very small for a nearly incompressible fluid, in which case the speed of sound is correspondingly large.

[13]Note that thermal expansion induces negligible corrections to Eq. (3.3) of order $\beta\Delta T \lesssim 10^{-2}$.

$(\Delta T, h)$ enter through the appropriate boundary conditions that will be specified below.[14]

The Boussinesq equations have an elementary static solution representing uniform heat conduction with $\mathbf{v} = \mathbf{0}$ and a linear temperature profile $T^0(z)$ given by Eq. (3.4). The corresponding pressure $p^0(z)$ follows from the modified Navier-Stokes equation (3.14) with the static linear temperature distribution $T^0(z)$. An elementary calculation gives

$$p^0(z) = p_0 + \rho_0 g \left(h - z \right) - \frac{\rho_0 g \beta \Delta T}{2h} \left(h - z \right)^2 \qquad (3.15)$$

where p_0 is the pressure at the upper surface. The first two terms are the usual linear pressure dependence for a uniform fluid, and the last is a quadratic correction that reflects the density variation induced by thermal expansion. As expected from the physics, the decreased density near the bottom acts to reduce the pressure there, but the correction is small, since $\beta \Delta T \lesssim 0.01$.

Given the static conductive solution of the Boussinesq equations discussed above, we now investigate variations around this configuration and write

$$
\begin{aligned}
\mathbf{v} &= \mathbf{v}(\mathbf{x}, t) \\
p &= p^0(z) + \delta p(\mathbf{x}, t) \\
T &= T^0(z) + \delta T(\mathbf{x}, t)
\end{aligned}
\qquad (3.16)
$$

Here $\mathbf{v}$, δp and δT are the deviations (for the moment, they need not be small, although we shall indeed consider the linearized small-amplitude behavior below).

Since $p^0(z)$ and $T^0(z)$ together obey the coupled Navier-Stokes and heat-conduction equations, substitution of the relations (3.16) into Eq. (3.14) readily yields

$$\frac{d\mathbf{v}}{dt} = \frac{\partial \mathbf{v}}{\partial t} + (\mathbf{v} \cdot \boldsymbol{\nabla}) \mathbf{v} = -\frac{1}{\rho_0} \boldsymbol{\nabla} \delta p + g \beta \hat{\mathbf{z}} \, \delta T + \nu \nabla^2 \mathbf{v} \qquad (3.17)$$

Correspondingly, the heat conduction equation (3.1) takes the form

$$\frac{\partial \delta T}{\partial t} - \mathbf{v} \cdot \hat{\mathbf{z}} \frac{\Delta T}{h} + \mathbf{v} \cdot \boldsymbol{\nabla} \, \delta T = \kappa \nabla^2 \, \delta T \qquad (3.18)$$

No assumption has yet been made about the velocity field $\mathbf{v}$, other than incompressible flow

$$\boldsymbol{\nabla} \cdot \mathbf{v} = 0 \qquad (3.19)$$

It is important to recognize that these equations still contain quadratic nonlinear convective terms involving $\mathbf{v}$ and δT. By construction, the static solution to these equations is now explicitly given by

$$\mathbf{v} = 0 \qquad \delta p = 0 \qquad \delta T = 0 \quad ; \text{ static solution} \qquad (3.20)$$

---

[14]For a different but essentially equivalent derivation of the same equations, see, for example, [La87a].

22

with the corresponding static expressions for the temperature and pressure given in Eqs. (3.4) and (3.15). In the next three sections, we shall investigate the stability of the conductive state, the transition to the convective state, and the nature of the ensuing convective state.

Linearized perturbation equations

The Boussinesq equations can now be *linearized* about the static solution. It is straightforward to verify that they then take the form

$$\frac{\partial \mathbf{v}}{\partial t} = -\boldsymbol{\nabla}\delta w + g\beta\hat{\mathbf{z}}\,\delta T + \nu\,\nabla^2\mathbf{v}$$

$$\frac{\partial \delta T}{\partial t} - \frac{\Delta T}{h}v_z = \kappa\nabla^2\delta T$$

$$\boldsymbol{\nabla}\cdot\mathbf{v} = 0 \qquad (3.21)$$

where, for convenience, we follow [La87b] and redefine the pressure variation as

$$\delta w \equiv \frac{\delta p}{\rho_0} \qquad (3.22)$$

We proceed to rewrite these equations in dimensionless form. Introduce the following two dimensionless ratios

$$R \equiv \frac{gh^3\beta\,\Delta T}{\nu\kappa} \quad ; \text{ Rayleigh number}$$

$$P \equiv \frac{\nu}{\kappa} \quad ; \text{ Prandtl number} \qquad (3.23)$$

The dimensions are readily verified starting from the dimensions of (here l is length and t is time)

$$[\nu] = \left[\frac{l^2}{t}\right] ; \qquad [\kappa] = \left[\frac{l^2}{t}\right] ; \qquad [\beta\Delta T] = [1] \qquad (3.24)$$

Introduce also the following units of length, time, temperature, and pressure

$$h = \text{unit of length}$$

$$\frac{h^2}{\nu} = \text{unit of time}$$

$$\frac{\nu\,\Delta T}{\kappa} = P\,\Delta T = \text{unit of temperature}$$

$$\frac{\rho_0\nu^2}{h^2} = \text{unit of pressure} \qquad (3.25)$$

In terms of these units, after some algebra, Eqs. (3.21) become

$$\frac{\partial \mathbf{v}}{\partial t} = -\boldsymbol{\nabla}\delta w + \nabla^2\mathbf{v} + R\hat{\mathbf{z}}\,\delta T$$

$$P\frac{\partial \delta T}{\partial t} = \nabla^2\,\delta T + v_z$$

$$\boldsymbol{\nabla}\cdot\mathbf{v} = 0 \qquad (3.26)$$

23

These are the *dimensionless, linearized* Boussinesq equations for the Rayleigh-Bénard problem posed through Eqs. (3.1)–(3.3) and Fig. 3.1. There are now five coupled differential equations that are *linear* in the five functions $(\mathbf{v}, \delta w, \delta T)$ describing variations around the static solution in Eqs. (3.4) and (3.15). The fluid is characterized by the two dimensionless parameters (R, P) in Eq. (3.23). The third equation above ($\nabla \cdot \mathbf{v} = 0$) is simply a constraint on the velocity field (similar to $\nabla \cdot \mathbf{B} = 0$ in Maxwell's equations). To define the problem, we must, of course, specify the boundary conditions, which we proceed to do.

Boundary conditions

Suppose one simply has rigid walls at the two surfaces $z = 0$ and $z = 1$ (dimensionless units) in Fig. 3.1. Then the normal component of the fluid velocity must vanish there. If the walls are of the "non-slip" variety, then the tangential components of the fluid velocity must also vanish there. Suppose, further, that these walls are externally maintained at a fixed temperature so that δT is forced to vanish on them. In this case, the surface boundary conditions are

$$
\begin{aligned}
\mathbf{v} &= 0 \\
\delta T &= 0 \qquad ; z = 0, 1
\end{aligned}
\tag{3.27}
$$

The first relation actually represents six boundary conditions since it implies that $(v_x, v_y, v_z) = 0$ on each of the two surfaces.

Now everywhere in the fluid, the condition of incompressibility requires that

$$
\nabla \cdot \mathbf{v} = \frac{\partial v_x}{\partial x} + \frac{\partial v_y}{\partial y} + \frac{\partial v_z}{\partial z} = 0
\tag{3.28}
$$

Since $v_x = v_y = 0$ for all (x, y) on the non-slip surface, one has $\partial v_x / \partial x = \partial v_y / \partial y = 0$ on each wall. Equation (3.28) then implies the additional condition on the rigid non-slip walls

$$
\frac{\partial v_z}{\partial z} = 0 \qquad ; z = 0, 1
\tag{3.29}
$$

In summary, for *rigid, non-slip surfaces with fixed temperatures* the boundary conditions are

$$
\begin{aligned}
v_z &= \frac{\partial v_z}{\partial z} = 0 \\
v_x &= v_y = 0 \\
\delta T &= 0 \qquad ; z = 0, 1
\end{aligned}
\tag{3.30}
$$

Equations (3.30) provide five boundary conditions on each fixed surface.

For mathematical convenience, it is often useful to imagine a *free surface* that can move laterally to eliminate any tangential stress but is still fixed in the z direction and held at a fixed temperature (as seen below, this slightly unphysical model allows a complete analytical solution of the Rayleigh-Bénard problem). The model still requires the boundary conditions that $v_z = 0$ and $\delta T = 0$ everywhere on the surfaces. As a consequence of the first relation, it follows immediately that $\partial v_z/\partial x = \partial v_z/\partial y = 0$ everywhere on these free surfaces.

For an incompressible viscous fluid, the stress tensor in Eq. (60.23b) takes the form

$$T_{ij} = \rho_0 v_i v_j + p\delta_{ij} - \eta \left(\frac{\partial v_i}{\partial x_j} + \frac{\partial v_j}{\partial x_i} \right) \tag{3.31}$$

where T_{ij} is the ith component of the force per unit area on a surface oriented along the jth direction [see Eq. (60.4)]. In the case of a viscous fluid, the term proportional to the viscosity η can lead to a tangential shear on the surface. For a free surface, the appropriate boundary condition is that the tangential stresses T_{xz} and T_{yz} must vanish, which ensures that the x and y components of the force must vanish on the bounding surfaces (here oriented in the z direction).

These arguments imply the following conditions on a free surface

$$v_z = \frac{\partial v_z}{\partial x} = 0$$

$$T_{xz} = -\eta \left(\frac{\partial v_x}{\partial z} + \frac{\partial v_z}{\partial x} \right) = 0 \tag{3.32}$$

A combination of the two conditions yields (a similar argument holds for T_{yz})

$$\frac{\partial v_x}{\partial z} = \frac{\partial v_y}{\partial z} = 0 \tag{3.33}$$

Now take $\partial/\partial z$ of Eq. (3.28) and interchange the order of partial derivatives. This procedure gives the final boundary condition on a free surface

$$\frac{\partial^2 v_z}{\partial z^2} = 0 \tag{3.34}$$

In summary, for *tangentially-free surfaces at fixed temperatures* the boundary conditions are

$$v_z = \frac{\partial^2 v_z}{\partial z^2} = 0$$

$$\frac{\partial v_x}{\partial z} = \frac{\partial v_y}{\partial z} = 0$$

$$\delta T = 0 \qquad\qquad ; z = 0, 1 \tag{3.35}$$

Equations (3.35) again yield five boundary conditions on each surface. It is, of course, also possible to have different boundary conditions, fixed or free, on each of the two surfaces.

4 Rayleigh-Bénard problem: linearized theory of convective instability

We next seek *normal-mode* solutions to the linearized Eqs. (3.26). The coupled unknown functions must have a common time dependence, which then cancels from the equations upon substitution. Instead of oscillations with a frequency ω (as found in Sec. 2), we here anticipate real exponential behavior and assume

$$
\begin{aligned}
\mathbf{v}(\mathbf{x}, t) &= \mathbf{v}(\mathbf{x})\, e^{\gamma t} \\
\delta w(\mathbf{x}, t) &= w(\mathbf{x})\, e^{\gamma t} \\
\delta T(\mathbf{x}, t) &= \tau(\mathbf{x})\, e^{\gamma t}
\end{aligned}
\tag{4.1}
$$

This set of coupled equations will be seen to yield a self-adjoint eigenvalue problem for the parameter γ, which can be proved to be *real* directly from the equations themselves. As expected from the behavior of other eigenvalue problems, there will be a set of normal modes, and the general solution can be obtained by linear superposition. From the discussion in Sec. 2, we know that a *negative* value of the time constant γ implies exponential decay of the deviation back to the static solution, while a *positive* value of γ implies exponential growth away from the static solution and indicates instability. We proceed to demonstrate that γ is real.

Proof that solutions are not oscillatory

After substitution of Eqs. (4.1) into Eqs. (3.26) and cancellation of the common factor $e^{\gamma t}$, they take the linear and time-independent form

$$
\begin{aligned}
\gamma\,\mathbf{v}(\mathbf{x}) &= -\boldsymbol{\nabla} w(\mathbf{x}) + \nabla^2 \mathbf{v}(\mathbf{x}) + R\tau(\mathbf{x})\,\hat{\mathbf{z}} \\
\gamma P \tau(\mathbf{x}) &= \nabla^2 \tau(\mathbf{x}) + v_z(\mathbf{x}) \\
\boldsymbol{\nabla} \cdot \mathbf{v}(\mathbf{x}) &= 0
\end{aligned}
\tag{4.2}
$$

The first two equations can be considered a generalized eigenvalue equation for a two-component vector with elements $[\mathbf{v}(\mathbf{x}), \tau(\mathbf{x})]$, since acting with the operators on the right hand side yields the eigenvalue γ times the same vector (apart from the term $-\boldsymbol{\nabla} w$ and an additional constant factor P in the second component). The last equation constrains the allowed velocity field $\mathbf{v}(\mathbf{x})$. Note that the amplitudes $[\mathbf{v}(\mathbf{x}), \tau(\mathbf{x})]$ will depend on the coordinate $\mathbf{x}$, which is similar to the general Sturm-Liouville eigenvalue problem studied in Secs. 40 and 41. Nevertheless, we are in a position to determine some important properties of γ [La87b].

Assume, for the purposes of this discussion, that the system in Fig. 3.1 is in a cavity with rigid, non-slip walls and that $\mathbf{v}$ thus vanishes on all the boundaries. Assume, in addition, that all of the walls are held at the appropriate temperature so that $\tau = 0$ on all the walls.

Now suppose that $(\gamma, \mathbf{v}, w, \tau)$ are complex. Dot the first of Eqs. (4.2) into

26

$\mathbf{v}^\star$ and integrate over the cavity volume

$$\gamma \int_{\text{vol}} d^3x\,|\mathbf{v}|^2 \;=\; -\int_{\text{vol}} d^3x\,\mathbf{v}^\star \cdot \boldsymbol{\nabla}w + \int_{\text{vol}} d^3x\,\mathbf{v}^\star \cdot \nabla^2\mathbf{v} + \int_{\text{vol}} d^3x\,R\tau v_z^\star \quad (4.3)$$

Next carry out the following steps:

1. Use the constraint $\boldsymbol{\nabla} \cdot \mathbf{v} = 0$ to write

$$\mathbf{v}^\star \cdot \boldsymbol{\nabla}w = \boldsymbol{\nabla} \cdot (w\mathbf{v}^\star) \qquad (4.4)$$

   The divergence theorem converts the resulting volume integral to a surface integral over the walls, where the integrand vanishes due to the boundary conditions;

2. For the same reason, one can write

$$\begin{aligned}
\mathbf{v}^\star \cdot \nabla^2\mathbf{v} &= -\mathbf{v}^\star \cdot (\boldsymbol{\nabla} \times \boldsymbol{\nabla} \times \mathbf{v}) \\
&= \boldsymbol{\nabla} \cdot [\mathbf{v}^\star \times (\boldsymbol{\nabla} \times \mathbf{v})] - |\boldsymbol{\nabla} \times \mathbf{v}|^2 \qquad (4.5)
\end{aligned}$$

   The integral of the first term can again be transformed into a surface integral over the walls, where it vanishes.

The result is

$$\gamma \int_{\text{vol}} d^3x\,|\mathbf{v}|^2 \;=\; \int_{\text{vol}} d^3x\,\left[-|\boldsymbol{\nabla} \times \mathbf{v}|^2 + R\tau v_z^\star\right] \qquad (4.6)$$

Note that the pressure w no longer appears.

Similarly, multiply the second of Eqs. (4.2) by $\tau^\star$

$$\gamma P \int_{\text{vol}} d^3x\,|\tau|^2 \;=\; \int_{\text{vol}} d^3x\,\tau^\star \nabla^2\tau + \int_{\text{vol}} d^3x\,\tau^\star v_z \qquad (4.7)$$

Write

$$\tau^\star \nabla^2\tau \;=\; \boldsymbol{\nabla} \cdot (\tau^\star \boldsymbol{\nabla}\tau) - |\boldsymbol{\nabla}\tau|^2 \qquad (4.8)$$

The divergence theorem transforms the integral of the first term into a surface integral, which vanishes because $\tau = 0$ there, giving the result

$$\gamma P \int_{\text{vol}} d^3x\,|\tau|^2 \;=\; \int_{\text{vol}} d^3x\,\left[-|\boldsymbol{\nabla}\tau|^2 + \tau^\star v_z\right] \qquad (4.9)$$

Subtract their complex conjugates from each of Eqs. (4.6) and (4.9)

$$\begin{aligned}
(\gamma - \gamma^\star) \int_{\text{vol}} d^3x\,|\mathbf{v}|^2 &= R \int_{\text{vol}} d^3x\,[\tau v_z^\star - \tau^\star v_z] \\
(\gamma - \gamma^\star)P \int_{\text{vol}} d^3x\,|\tau|^2 &= -\int_{\text{vol}} d^3x\,[\tau v_z^\star - \tau^\star v_z] \qquad (4.10)
\end{aligned}$$

27

Now multiply the second equation by R and add to the first

$$(\gamma - \gamma^*) \int_{\text{vol}} d^3x \, \left[|\mathbf{v}|^2 + RP|\tau|^2 \right] = 0 \qquad (4.11)$$

Since the integral is positive definite, one concludes that

$$\gamma = \gamma^* \qquad (4.12)$$

Hence γ is real.[15]

Since γ is real, one can also take $(\mathbf{v}, \tau)$ real in Eqs. (4.6) and (4.9).[16] Multiply Eq. (4.9) by R and add it to Eq. (4.6):

$$\gamma \int_{\text{vol}} d^3x \, \left[v^2 + RP\tau^2 \right] = \int_{\text{vol}} d^3x \, \left[-(\nabla \times \mathbf{v})^2 - R(\nabla \tau)^2 + 2R\tau v_z \right] \qquad (4.13)$$

Equivalently, we can solve for γ

$$\gamma = \frac{I_1}{I_2}$$

$$I_1 = \int_{\text{vol}} d^3x \, \left[-(\nabla \times \mathbf{v})^2 - R(\nabla \tau)^2 + 2R\tau \, v_z \right]$$

$$I_2 = \int_{\text{vol}} d^3x \, \left[v^2 + RP\,\tau^2 \right] \qquad (4.14)$$

which provides an explicit expression for γ in terms of $[\mathbf{v}(\mathbf{x}), \tau(\mathbf{x})]$. As noted above, the pressure variation $w(\mathbf{x})$ no longer appears in these expressions.

It is instructive to observe that the above relation serves as a variational principle for γ. Consider the variations $(\delta \mathbf{v}, \delta \tau)$ about the solutions $(\mathbf{v}, \tau)$ and require that γ be stationary under these variations

$$\delta\gamma = \frac{\delta I_1}{I_2} - \frac{I_1}{I_2^2} \delta I_2 = \frac{\delta I_1 - \gamma \delta I_2}{I_2} = 0 \qquad (4.15)$$

Now the components of $\delta \mathbf{v}$ are not all independent since there is a constraint on the velocity field

$$\nabla \cdot (\mathbf{v} + \delta \mathbf{v}) = \nabla \cdot (\delta \mathbf{v}) = 0 \qquad (4.16)$$

This constraint can be incorporated into the variational principal through a Lagrange multiplier.[17] Multiply Eq. (4.16) by an arbitrary function $w(\mathbf{x})$ and integrate over the volume; the result is still zero

$$\int_{\text{vol}} d^3x \, w \nabla \cdot (\delta \mathbf{v}) = 0 \qquad (4.17)$$

---

[15] A similar result is established for free surfaces in [Ch81c].

[16] Just take Re of Eqs. (4.2).

[17] Alternatively, if one does not use a Lagrange multiplier to impose the constraint, then the only thing one can deduce from the corresponding Eq. (4.20) is that $R\tau\hat{\mathbf{z}} + \nabla^2\mathbf{v} - \gamma\mathbf{v} = \nabla w$ where w is simply some function of $\mathbf{x}$, for then Eqs. (4.18) and (4.17) imply that the overall variation $\delta\gamma$ still vanishes.

Next integrate by parts [analogous to Eq. (4.4)]. The total divergence can be integrated out to the surface where it vanishes due to our boundary conditions, yielding

$$-\int_{\text{vol}} d^3x \, (\delta \mathbf{v}) \cdot \boldsymbol{\nabla} w = 0 \qquad (4.18)$$

Now add Eq. (4.18) to the numerator in Eq. (4.15).

Consider the variation $\delta\tau(\mathbf{x})$. With the aid of some algebra and judicious use of the divergence theorem and boundary conditions, Eq. (4.15) then gives

$$\int_{\text{vol}} d^3x \, \left[\nabla^2 \tau + v_z - P\gamma\tau \right] \delta\tau = 0 \qquad (4.19)$$

If this expression is required to vanish for all $\delta\tau(\mathbf{x})$, one reproduces the second of Eqs. (4.2). Next, consider the variation $\delta\mathbf{v}(\mathbf{x})$. Somewhat more extensive algebra yields

$$\int_{\text{vol}} d^3x \, \left[-\boldsymbol{\nabla} w + R\tau\hat{\mathbf{z}} + \nabla^2 \mathbf{v} - \gamma\mathbf{v} \right] \cdot \delta\mathbf{v} = 0 \qquad (4.20)$$

If this expression is now required to vanish for *all* $\delta\mathbf{v}(\mathbf{x})$, then the first of Eqs. (4.2) is reproduced. It is still necessary to impose the third of Eqs. (4.2) as a constraint

$$\boldsymbol{\nabla} \cdot \mathbf{v} = 0 \qquad (4.21)$$

The presence of the Lagrange function $w(\mathbf{x})$ gives us the flexibility to do so. We learn from this exercise that the only role of the pressure term is to guarantee the constraint of incompressibility in Eq. (4.21)!

We observe from Eq. (4.14) that if R is small, then both I_1 and γ are negative. From Eqs. (4.1) and our previous discussion, this indicates that all perturbations decay and the static, conductive solution is stable. As R grows, there will be a value R_c at which I_1 and γ first vanish. For values of R greater than R_c, γ will become positive, indicating that at least some perturbations will grow exponentially. Hence the conductive state of the system will now be unstable, and a transition to convection will set in. The condition $\gamma = 0$ determines the critical value R_c.

Vorticity and the eigenvalue equation

One important concept in hydrodynamics is the *vorticity* $\boldsymbol{\zeta}$, defined as

$$\boldsymbol{\zeta} \equiv \boldsymbol{\nabla} \times \mathbf{v} \qquad (4.22)$$

In this context, Kelvin's circulation theorem in a nonviscous fluid, stated in Eq. (48.61), can be interpreted as a conservation law for the vorticity enclosed in any closed path that moves with the fluid. Equivalent differential forms of the same physics are studied in Probs. 9.5 and 9.6.

The situation is more subtle in a viscous fluid, for Prob. 12.2 shows that the vorticity in this case obeys a generalized diffusion equation, with the kinematic viscosity ν as the characteristic diffusion constant. More generally, the vorticity will be seen to play an important role in the current Rayleigh-Bénard problem.

Take the curl of the Navier-Stokes equation—the first of Eqs. (4.2)

$$\gamma(\nabla \times \mathbf{v}) = \nabla^2(\nabla \times \mathbf{v}) + R\nabla \times (\tau\hat{\mathbf{z}})$$

$$\text{or} \qquad \gamma\zeta = \nabla^2\zeta + R\nabla\tau \times \hat{\mathbf{z}} \tag{4.23}$$

Since $\nabla \times (\nabla w) = 0$, the pressure has disappeared from this relation.

It is valuable to repeat this procedure, namely take the curl of Eq. (4.23), and use

$$\nabla \times (\nabla \times \mathbf{v}) = \nabla(\nabla \cdot \mathbf{v}) - \nabla^2\mathbf{v} = -\nabla^2\mathbf{v}$$

$$\nabla \times [\nabla \times (\tau\hat{\mathbf{z}})] = \nabla\left(\frac{\partial\tau}{\partial z}\right) - \nabla^2(\tau\hat{\mathbf{z}}) \tag{4.24}$$

Hence

$$\gamma\nabla^2\mathbf{v} = \nabla^4\mathbf{v} - R\nabla\left(\frac{\partial\tau}{\partial z}\right) + R\nabla^2(\tau\hat{\mathbf{z}}) \tag{4.25}$$

Now take the z component of this relation. The result is that the first two of Eqs. (4.2) can be recast in the still-exact form

$$\gamma\nabla^2 v_z = \nabla^4 v_z + R\nabla_\perp^2\tau$$

$$\gamma P\tau = \nabla^2\tau + v_z \tag{4.26}$$

Here

$$\nabla_\perp^2 \equiv \frac{\partial^2}{\partial x^2} + \frac{\partial^2}{\partial y^2} = \nabla^2 - \frac{\partial^2}{\partial z^2} \tag{4.27}$$

is the transverse part of the Laplacian. Note that the problem has been reduced to two coupled equations for (v_z, τ) together with the previously discussed boundary conditions at the surfaces. The pressure has disappeared, but the equations are now of higher order in spatial derivatives.

With the development in Sec. 2 as a guide, let us examine normal-mode excitations in the x direction. Translational invariance in the x direction suggests that plane waves $\propto e^{iqx}$ are the appropriate eigenfunctions, since $\partial/\partial x$ then yields iq when applied to such functions. Thus we seek solutions in the form

$$v_z(x, z) = v_z(z)e^{iqx}$$

$$\tau(x, z) = \tau(z)e^{iqx} \tag{4.28}$$

Acting on these functions, one has

$$\nabla^2 \rightarrow \frac{d^2}{dz^2} - q^2$$

$$\nabla_\perp^2 \rightarrow -q^2 \tag{4.29}$$

30

Thus, upon substitution of Eqs. (4.28) and cancellation of the e^{iqx}, Eqs. (4.26) become ordinary differential equations in z

$$\left(\frac{d^2}{dz^2} - q^2\right)\left(\frac{d^2}{dz^2} - q^2 - \gamma\right)v_z(z) = R q^2 \tau(z)$$

$$\left(\frac{d^2}{dz^2} - q^2 - P\gamma\right)\tau(z) = -v_z(z) \qquad (4.30)$$

This set of equations still represents a linear eigenvalue problem for the real quantity $\gamma(P, q, R)$. These equations must be combined with the boundary conditions in either Eqs. (3.30) or (3.35)

Rigid walls	Free walls	
$\tau = 0$	$\tau = 0$	
$v_z = 0$	$v_z = 0$	
$\dfrac{dv_z}{dz} = 0$	$\dfrac{d^2 v_z}{dz^2} = 0$	(4.31)

The pressure w and the transverse velocities (v_x, v_y) have now been eliminated from the problem. Given v_z, the latter quantities can be reconstructed from $\nabla \cdot \mathbf{v} = 0$ and the boundary conditions (see later). The pressure, which is rarely needed, can then be reconstructed from the Navier-Stokes equation.

Since $\gamma = 0$ marks the onset of instability, we concentrate on this value, so that Eqs. (4.30) then take the simpler form

$$\left(\frac{d^2}{dz^2} - q^2\right)^2 v_z(z) = R q^2 \tau(z)$$

$$\left(\frac{d^2}{dz^2} - q^2\right)\tau(z) = -v_z(z) \qquad ; \gamma = 0 \qquad (4.32)$$

Furthermore, τ can be eliminated by letting $(d^2/dz^2 - q^2)$ act on the first equation and substituting the second

$$\left(\frac{d^2}{dz^2} - q^2\right)^3 v_z(z) = -R q^2 v_z(z) \qquad (4.33)$$

When combined with appropriate boundary conditions, this equation will determine the critical Rayleigh number R_c for the onset of convection. Note that the Prandtl number P no longer appears and thus does not affect the critical value of the Rayleigh number R_c.

This equation for v_z is now a sixth-order differential equation in z. In addition to the boundary conditions in Eq. (4.31), there is an additional boundary condition on v_z arising from the condition that $\tau = 0$ on both bounding surfaces. From the first of Eqs. (4.32), we require

$$\left(\frac{d^2}{dz^2} - q^2\right)^2 v_z(z) = 0 \qquad ; \text{ on boundaries} \qquad (4.34)$$

31

along with $v_z = 0$ and either $dv_z/dz = 0$ or $d^2v_z/dz^2 = 0$.

For general (q, R), one cannot find a solution to the differential equation that satisfies all six boundary conditions. If q is specified, Eq. (4.33) becomes an eigenvalue problem for R with a discrete set of eigenvalues $R_n(q)$ where $n = 1, 2, \cdots$. We select the lowest eigenvalue $R_1(q)$ and then seek its minimum with respect to q, which will occur at a critical wave number q_c. The result of this procedure will yield

$$R_1(q_c) = R_c \qquad ; \text{ the critical Rayleigh number}$$
$$q_c \qquad ; \text{ the critical wave number} \qquad (4.35)$$

These are the values at which the static, conductive state in the Boussinesq approximation to the Rayleigh-Bénard problem first undergoes the convective instability. We proceed to discuss the solution to the problem as posed here.

Free-free boundary conditions: exact solution

Remarkably, it is possible to find an *exact elementary solution* in the case of tangentially-free bounding surfaces. Although this problem is slightly unphysical in most cases, it is much simpler than the task of solving the problem with rigid boundary conditions. The principal effect of the more realistic rigid boundaries is to modify the numerical details, leaving the basic physical picture unchanged.[18]

We re-summarize the basic equations for the perturbation around the static, conductive configuration in Fig. 3.1. The z component of the fluid velocity must satisfy

$$\left(\frac{d^2}{dz^2} - q^2 \right)^3 v_z(z) = -R q^2 v_z(z) \qquad (4.36)$$

The corresponding free boundary conditions are

$$v_z = 0$$
$$\frac{d^2 v_z}{dz^2} = 0$$
$$\left(\frac{d^2}{dz^2} - q^2 \right)^2 v_z(z) = 0 \qquad ; z = 0, 1 \qquad (4.37)$$

Take as solution

$$v_z(z) = f_n(z) = \sin(n\pi z) \qquad ; n = 1, 2, \cdots \qquad (4.38)$$

On the boundaries, one can readily check that

$$f_n(z) = 0$$
$$\frac{d^2 f_n(z)}{dz^2} = -n^2\pi^2 f_n(z) = 0$$
$$\left(\frac{d^2}{dz^2} - q^2 \right) f_n(z) = 0 \qquad ; z = 0, 1 \qquad (4.39)$$

---

[18]We return to this question in the next subsection.

In fact, all even derivatives of $f_n(z)$ vanish on the two boundaries. Substituting this solution into Eq. (4.36), one finds

$$(-1)^3(n^2\pi^2 + q^2)^3 f_n(z) = -Rq^2 f_n(z) \tag{4.40}$$

Thus Eq. (4.36) will be satisfied in the fluid interior if the Rayleigh number R is one of the eigenvalues defined by

$$R_n(q) = \frac{(n^2\pi^2 + q^2)^3}{q^2} \qquad ; n = 1, 2, \cdots \tag{4.41}$$

Recall from Eq. (4.28) that q is the wave number of the normal mode in the x direction, which we can take to lie in the plane of Fig. 3.1.

Let us focus on the lowest eigenvalue $R_1(q) = (\pi^2 + q^2)^3/q^2$. It is clear by inspection that $R_1(q)$ diverges for both large and small q^2, so that it has a minimum at some intermediate finite value q_c^2. To minimize this expression, set its derivative equal to zero

$$\frac{dR_1}{dq^2} = \frac{3(\pi^2 + q^2)^2}{q^2} - \frac{(\pi^2 + q^2)^3}{q^4} = 0$$

$$q_c^2 = \frac{\pi^2}{2} \tag{4.42}$$

In *summary*, for the mode satisfying free-free boundary conditions with minimum Rayleigh number $R_1(q_c)$

$$
\begin{aligned}
v_z(z) &= f_1(z) = \sin(\pi z) \\
v_z(x, z) &= \cos(q_c x)\, v_z(z) = \cos(q_c x)\sin(\pi z) \\
q_c &= \frac{\pi}{\sqrt{2}} = 2.221\cdots \\
\lambda_c &= \frac{2\pi}{q_c} = 2\sqrt{2} = 2.828\cdots \\
R_c &= R_1(q_c) = \frac{27\pi^4}{4} = 657.5\cdots
\end{aligned}
\tag{4.43}
$$

where the real part is now made explicit. Recall that lengths here are measured in units of h, and the dimensionless Rayleigh number $R = gh^3\beta\Delta T/(\nu\kappa)$ is defined in Eq. (3.23). For $R < R_c$, all modes decay and the stationary conductive state is stable. For $R = R_c$ the mode with $q = q_c$ just becomes unstable, and for $R > R_c$, this convective mode with wavelength λ_c will be the first to grow exponentially with time.[19]

The corresponding temperature perturbation can be obtained from the second of Eqs. (4.32)

$$\tau(x, z) = \frac{\cos(q_c x)\sin(\pi z)}{\pi^2 + q_c^2} \tag{4.44}$$

---

[19]As shown in the problems, an explicit solution of Eq. (4.30) demonstrates this behavior in detail.

The transverse velocity $v_x(x, z)$ for this solution (there is no v_y) can be determined from the relation

$$\nabla \cdot \mathbf{v} = \frac{\partial v_x}{\partial x} + \frac{\partial v_z}{\partial z} = 0$$

$$\frac{\partial v_x}{\partial x} = -\frac{\partial v_z}{\partial z} = -\pi \cos(q_c\, x) \cos(\pi z)$$

$$v_x(x, z) = -\frac{\pi}{q_c} \sin(q_c\, x) \cos(\pi z) \qquad (4.45)$$

The solution evidently satisfies the tangential boundary condition in Eqs. (3.35)

$$\frac{\partial v_x}{\partial z} = 0 \qquad ; z = 0, 1 \qquad (4.46)$$

It is instructive to examine some of the properties of this first unstable mode about the static conductive state in Fig. 3.1

$$v_z(x, z) = v_0 \cos(q_c\, x) \sin(\pi z)$$

$$v_x(x, z) = -\frac{v_0 \pi}{q_c} \sin(q_c\, x) \cos(\pi z)$$

$$\tau(x, z) = \frac{v_0}{\pi^2 + q_c^2} \cos(q_c\, x) \sin(\pi z) \qquad (4.47)$$

Here v_0 is some small, first-order amplitude, and the critical wave number $q_c = \pi/\sqrt{2}$, wavelength $\lambda_c = 2\pi/q_c = 2\sqrt{2}$, and Rayleigh number $R_c = 27\pi^4/4$ are given in Eqs. (4.43). A few key values of the vertical velocity are

$$
\begin{aligned}
v_z(x, z) &= 0 & &; z = 0, 1 \\
&= 0 & &; z = 1/2 \ ; x = \lambda_c/4,\ 3\lambda_c/4,\ \cdots \\
&= v_0 & &; z = 1/2 \ ; x = 0,\ \lambda_c,\ 2\lambda_c,\ \cdots \\
&= -v_0 & &; z = 1/2 \ ; x = \lambda_c/2,\ 3\lambda_c/2,\ \cdots \qquad (4.48)
\end{aligned}
$$

Some corresponding values of the transverse velocity are

$$
\begin{aligned}
v_x(x, z) &= -\frac{v_0 \pi}{q_c} \cos(\pi z) & &; x = \lambda_c/4 \\
&= 0 & &; x = \lambda_c/2 \qquad (4.49)
\end{aligned}
$$

Note that $\mathbf{v}^2$ everywhere in the interior of the fluid satisfies

$$
\begin{aligned}
\mathbf{v}^2 &= v_x^2 + v_z^2 \\
&= v_0^2 \left[2\sin^2(q_c\, x) \cos^2(\pi z) + \cos^2(q_c\, x) \sin^2(\pi z) \right] \\
&\geq 0 \qquad (4.50)
\end{aligned}
$$

It vanishes at the center and corners of each half cell. Since $\nabla \cdot \mathbf{v} = 0$, the flow lines in the fluid will be continuous and closed, making it relatively easy to sketch the streamlines. This situation is illustrated in Fig. 4.1.

34

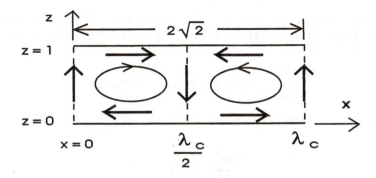

Fig. 4.1. Sketch of the flow pattern for the first unstable mode of the static conductive configuration in Fig. 3.1. It is a convective mode, with cells containing pairs of opposite convective rolls whose size in dimensionless units is noted in the figure. All lengths here are measured in units of h.

The vorticity lies along $\hat{y}$

$$\zeta = \nabla \times v$$
$$= \zeta \hat{y} \tag{4.51}$$

and the definition of the curl yields

$$\zeta = \frac{\partial v_x}{\partial z} - \frac{\partial v_z}{\partial x}$$
$$= \frac{\pi^2}{q_c} v_0 \sin(q_c x) \sin(\pi z) + q_c v_0 \sin(q_c x) \sin(\pi z)$$
$$= \frac{v_0}{q_c}(\pi^2 + q_c^2) \sin(q_c x) \sin(\pi z) \tag{4.53}$$

This quantity takes the values

$$\zeta = \frac{v_0}{q_c}(\pi^2 + q_c^2) \qquad ; z = 1/2 \qquad ; x = \lambda_c/4, 5\lambda_c/4 \cdots$$
$$= -\frac{v_0}{q_c}(\pi^2 + q_c^2) \qquad ; z = 1/2 \qquad ; x = 3\lambda_c/4, 7\lambda_c/4 \cdots \tag{4.54}$$

Its magnitude is maximized at the center of each cell, and it reverses sign in going from cell to cell, as one might have surmised from Fig. 4.1 (note that $\hat{y}$ points into the plane).

Rigid-rigid boundary conditions: sketch of exact solution

We are able to find a simple analytic solution for free-free boundary conditions. In this case, the fluid has a nonzero tangential velocity at the walls (Fig. 4.1), and the surface simply accommodates this velocity. In the case of rigid-rigid boundary conditions, however, the tangential components of the velocity must vanish at the surface. Although the fluid tries to accommodate the convective rolls illustrated in Fig. 4.1, the solution evidently becomes more complicated. One is faced with solving a linear, sixth-order differential equation with constant coefficients

$$\left(\frac{d^2}{dz^2} - q^2\right)^3 v_z(z) = -Rq^2 v_z(z) \tag{4.55}$$

subject to the six boundary conditions

$$
\begin{aligned}
v_z &= 0 \\
\frac{dv_z}{dz} &= 0 \\
\left(\frac{d^2}{dz^2} - q^2\right)^2 v_z(z) &= 0 \qquad ; z = 0, 1
\end{aligned}
\tag{4.56}
$$

on each horizontal bounding surface.

The lowest eigenvalue $R_1(q)$ will be associated with the solution that is even around the central plane. The equations can then be solved by assuming

$$
\begin{aligned}
v_z &= e^{\mu z} \\
(\mu^2 - q^2)^3 &= -Rq^2
\end{aligned}
\tag{4.57}
$$

where $\mu = \mu(q, R)$. This gives a cubic equation for μ^2 with six complex roots $(\pm\mu_1, \pm\mu_2, \pm\mu_3)$, each of which depends on R and q. Based on the even symmetry around $z = 1/2$, the appropriate linear combinations are $\cosh[\mu_i(z - 1/2)]$ for $i = 1, 2, 3$. Linear combinations of these three solutions must be chosen to match the boundary conditions. This procedure involves some tedious algebra and numerical work that will not be reproduced here. The result [Ch81c] is

Rigid-rigid	Free-free	
$R_c = 1708$	$R_c = 657.5$	
$q_c = 3.117$	$q_c = 2.221$	(4.58)

where we compare with the numbers obtained above in Eq. (4.43) in the free-free case. In [Ch81c] an elegant variational principle is developed for R_c for the rigid boundary conditions; a single-term approximation yields a numerical value that exceeds the above value by $\approx 1/2\%$.

We have only considered the case of convection rolls that are periodic along one direction. More generally, the linearized Boussinesq equations have other more complicated solutions that are periodic in the plane with a regular lattice of convection cells [Ch81c] (in fact, Bénard observed hexagonal cells, which occur quite frequently; see, for example, Fig. 1 in [Ch81]). To determine which of these various patterns actually appears in any particular situation requires the inclusion of the leading nonlinear terms that are omitted in the present approximation.

Instead of pursuing this matter, however, we instead turn to more general questions. So far we have shown that $R = R_c$ characterizes the onset of instability from the static conductive configuration to the convective rolls, working in the linearized Boussinesq approximation to the Rayleigh-Bénard problem. What happens when R increases beyond R_c, so that $(R - R_c)/R_c$ is not necessarily small? We must then attack the full nonlinear Boussinesq Eqs. (3.17), (3.18), and (3.19), which provide five coupled, nonlinear equations for the five quantities $(\mathbf{v}, \delta p, \delta T)$ representing variations about the static solution. We proceed to discuss this problem.

5   Rayleigh-Bénard problem: expansion in Fourier modes

For $R > R_c$ one observes a rich set of phenomena [Bo00]. The linearized stability analysis of Sec. 4 only identifies the onset of convective instability at $R = R_c$ and the critical convective wave number $q = q_c$. In contrast, it does not allow us to follow the time evolution of that mode for $R > R_c$. To study the behavior for finite $(R - R_c)/R_c$, we must extend the analysis to the full nonlinear case and include convection.

We observe that the nonlinear Boussinesq Eqs. (3.17), (3.18), and (3.19) can be recovered from the linearized versions through the replacement of the partial time derivative by the hydrodynamic derivative

$$\frac{\partial}{\partial t} \rightarrow \frac{\partial}{\partial t} + \mathbf{v} \cdot \nabla \tag{5.1}$$

Both of these terms have dimensions $[t^{-1}]$; hence the dimensionless set of Eqs. (3.17), (3.18), and (3.19) takes the form [compare Eqs. (3.26)]

$$\frac{\partial \mathbf{v}}{\partial t} + (\mathbf{v} \cdot \nabla)\mathbf{v} = -\nabla \delta w + \nabla^2 \mathbf{v} + R\hat{\mathbf{z}}\,\delta T$$

$$P\left(\frac{\partial \delta T}{\partial t} + \mathbf{v} \cdot \nabla \delta T\right) = \nabla^2 \delta T + v_z$$

$$\nabla \cdot \mathbf{v} = 0 \tag{5.2}$$

Here the units are those of Eqs. (3.25), and the quantities $(\mathbf{v}, \delta w, \delta T)$ represent deviations from the stationary values in Eqs. (3.4) and (3.15) [note Eq. (3.22)].

Finding the general solution to a set of coupled, nonlinear, partial differential equations presents a formidable problem, even with modern computing techniques. One can use various analytical methods, such as assuming $(R - R_c)/R_c \equiv \varepsilon$ is small and expanding in $(v_z, v_x) \propto \sqrt{\varepsilon}$ to work above R_c. Here we describe another approach, which permits an extension to larger R.

The previous analysis for the critical mode in the linearized case with free-free boundary conditions identified the solution

$$v_z(x, z) = v_0 \cos(qx) \sin(\pi z)$$

$$v_x(x, z) = -\frac{\pi}{q} v_0 \sin(qx) \cos(\pi z)$$

$$\tau(x, z) = \frac{v_0}{\pi^2 + q^2} \cos(qx) \sin(\pi z) \qquad ; q = q_c \tag{5.3}$$

This solution is just one possible Fourier amplitude that is periodic in x with period $\lambda = 2\pi/q$ and satisfies the free-free boundary conditions in z. It suggests an expansion of the full solution to the coupled problem as a double Fourier series satisfying the same conditions[20]

$$
\begin{aligned}
v_z(x, z, t) &= \sum_m \sum_n A_{m,n}(t) \cos{(mqx)} \sin{(n\pi z)} \\
v_x(x, z, t) &= \sum_m \sum_n B_{m,n}(t) \sin{(mqx)} \cos{(n\pi z)} \\
\delta T(x, z, t) &= \sum_m \sum_n C_{m,n}(t) \cos{(mqx)} \sin{(n\pi z)}
\end{aligned}
\qquad (5.4)
$$

Equations (5.2) also depend on the pressure $w(x, z, t)$; however, we know from the previous discussion that the pressure can be eliminated by taking the curl of the Navier-Stokes equation and working with the vorticity. Substitution of Eqs. (5.4) into the differential equations will then lead to an infinite set of coupled, nonlinear, *ordinary* differential equations in time for the Fourier coefficients. If this set is truncated for some reason, it reduces to a finite set of coupled equations that still exhibits the same nonlinearity. Such equations are readily integrated on PC's with currently available mathematics packages.

We here consider a very simple truncated set of such equations. Start from the velocity field of the critical lowest mode, with a time-dependent coefficient $u(t)$

$$
\begin{aligned}
v_z(x, z, t) &= u(t) \cos{(qx)} \sin{(\pi z)} \\
v_x(x, z, t) &= -\frac{\pi}{q} u(t) \sin{(qx)} \cos{(\pi z)}
\end{aligned}
\qquad (5.5)
$$

These components satisfy the condition

$$
\boldsymbol{\nabla} \cdot \mathbf{v} = \frac{\partial v_x}{\partial x} + \frac{\partial v_z}{\partial z} = 0
\qquad (5.6)
$$

The critical wavelength in the linearized case with free-free boundary conditions was found to be

$$
q = q_c \qquad ; q_c = \frac{\pi}{\sqrt{2}} \qquad ; \lambda_c = 2\sqrt{2}
\qquad (5.7)
$$

What about the temperature variation $\delta T(x, z, t)$? In the linearized theory, we had $\delta T \propto v_z$ from Eq. (4.44), so start with

$$
\delta T(x, z, t) = \tau_1(t) \cos{(qx)} \sin{(\pi z)}
\qquad (5.8)
$$

Now recall the Nusselt number in Eq. (3.7), which measures the total heat flux between the surfaces in Fig. 3.1. In the discussion in Sec. 3, we observed that

---

[20]In principle, the set of amplitudes can also include v_y, and they can also depend on y. We here ignore this possibility, although certain instabilities indeed require such additional flexibility [Mc75, Cr93].

the Nusselt number will show a break in slope at the critical value of R_c, where the convective roll with wavenumber q_c first forms. For $R < R_c$, the heat flux takes place by conduction and $N = 1$. For $R > R_c$, the heat flux also takes place by convection and $N > 1$, increasing with R (Fig. 5.1).

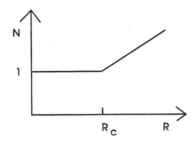

Fig. 5.1. Sketch of the Nussselt number in Eq. (3.7), which measures the heat flux between the surfaces in Fig. 3.1, as a function of the Rayleigh number R. See Fig. 14 in [Ch81] for typical experimental data with various fluids and non-slip surfaces, which give the experimental value $R_c = 1700 \pm 51$.

In order to describe convective heat transfer for $R > R_c$, it will turn out that we must add at least one more term to the temperature variation. In the last of Eqs. (5.4), we keep the simplest additional Fourier mode that suffices for us, that with $(m, n) = (0, 2)$ (the particular choice is justified below). Thus we look for a solution of the form

$$\delta T(x, z, t) = \tau_1(t) \cos(qx) \sin(\pi z) - \tau_2(t) \sin(2\pi z) \qquad (5.9)$$

To understand the presence of the additional term [Mc75], go back to the heat flow equation and rewrite it using the fact that $\boldsymbol{\nabla} \cdot \mathbf{v} = 0$

$$\frac{\partial T}{\partial t} + \mathbf{v} \cdot \boldsymbol{\nabla} T = \kappa \nabla^2 T$$

$$\frac{\partial T}{\partial t} + \boldsymbol{\nabla} \cdot (\mathbf{v} T - \kappa \boldsymbol{\nabla} T) = 0$$

$$\frac{\partial T}{\partial t} + \boldsymbol{\nabla}_\perp \cdot (\mathbf{v} T - \kappa \boldsymbol{\nabla}_\perp T) + \frac{\partial}{\partial z}\left(v_z T - \kappa \frac{\partial T}{\partial z}\right) = 0 \qquad (5.10)$$

Here $\boldsymbol{\nabla}_\perp$ is the transverse gradient. We temporarily revert to dimensional units and write for the perturbation in the temperature

$$\delta T(x, z, t) = \tau_1(t) \cos(qx) \sin\left(\frac{\pi z}{h}\right) - \tau_2(t) \sin\left(\frac{2\pi z}{h}\right) \qquad (5.11)$$

where (τ_1, τ_2) are expressed in absolute temperature °K.

As in the linear case, we seek a solution that is periodic in x (and independent of y). Introduce an average over the unit cell

$$\langle \mathbf{F}(x, z) \rangle \equiv \frac{1}{\lambda} \int_{\text{cell}} dx \, \mathbf{F}(x, z) \qquad (5.12)$$

39

where the wavelength λ is the spatial period of the cell structure. Since $\mathbf{F}(x, z)$ is periodic in x, we have

$$\langle \boldsymbol{\nabla}_\perp \cdot \mathbf{F}(x, z) \rangle = \left\langle \frac{\partial F_x(x, z)}{\partial x} \right\rangle = 0 \tag{5.13}$$

Here the transverse divergence is integrated out to the surface of the cell, and the contributions on opposite sides of the cell cancel because of the periodic boundary conditions.

Now take the cell average of Eq. (5.10), and set $\partial \langle T \rangle / \partial t = 0$ for the steady state:

$$\frac{\partial}{\partial z} \left(\langle v_z T \rangle - \kappa \frac{\partial \langle T \rangle}{\partial z} \right) = 0$$

$$\Rightarrow \quad \langle v_z T \rangle - \kappa \frac{\partial}{\partial z} \langle T \rangle = \text{constant} \qquad ; \text{steady state} \tag{5.14}$$

The left hand side of this last relation is the mean total heat flux through the cell in the z direction

$$-\kappa \frac{\partial \langle T \rangle}{\partial z} + \langle v_z T \rangle = \frac{1}{\rho c_p} \langle (\mathbf{j}_h^{\text{tot}})_z \rangle \tag{5.15}$$

For the configuration in Fig. 3.1, the temperature is given by

$$T = T_0 + \frac{\Delta T}{h}(h - z) + \delta T \tag{5.16}$$

Thus one has

$$\frac{1}{\rho c_p} \langle (\mathbf{j}_h^{\text{tot}})_z \rangle = \kappa \left(\frac{\Delta T}{h} - \frac{\partial \langle \delta T \rangle}{\partial z} \right) + \langle v_z T \rangle \tag{5.17}$$

In the steady state, Eq. (5.14) tells us that this quantity is a constant independent of z. Now substitute the cell-average of Eq. (5.11), noting that $\langle \cos(qx) \rangle = 0$, and evaluate the whole expression at both $z = (0, h)$ where $v_z = 0$. This procedure gives

$$\left\langle (\mathbf{j}_h^{\text{tot}})_z \right\rangle_{z=0} = \left\langle (\mathbf{j}_h^{\text{tot}})_z \right\rangle_{z=h} = k_{\text{th}} \left(\frac{\Delta T}{h} + \frac{2\pi}{h} \tau_2 \right) \tag{5.18}$$

The first term on the right hand side is the conductive heat flux identified previously in Sec. 3, which yields $N = 1$. The second term now provides a description of convection with $N > 1$. Even though there is no vertical fluid velocity at the bounding surfaces, the fluid flow throughout the cell transports heat away from the bottom surface and toward the upper surface; in effect, it increases the temperature gradient on each of them. Thus the second term in Eq. (5.9) serves the purpose for which it was introduced. Note that the additional term in Eq. (5.9) must

- have the form of one of the additional Fourier modes whose cell-average does not vanish (this rules out terms with $m \neq 0$);

- vanish on the two bounding surfaces;

- give the same stationary convective heat flux out of the bottom surface as into the top surface.

Let us then return to the dimensionless, coupled, nonlinear Eqs. (5.2) and look for solutions of the form

$$
\begin{aligned}
v_z(x, z, t) &= u(t)\cos(qx)\sin(\pi z) \\
v_x(x, z, t) &= -\frac{\pi}{q}u(t)\sin(qx)\cos(\pi z) \\
\delta T(x, z, t) &= \tau_1(t)\cos(qx)\sin(\pi z) - \tau_2(t)\sin(2\pi z)
\end{aligned}
\tag{5.19}
$$

Consider first the heat equation, the second of Eqs. (5.2), and substitute the above expression for $\delta T(x, z, t)$. Differentiation with respect to t yields

$$
P\frac{\partial \delta T}{\partial t} = P[\dot{\tau}_1\cos(qx)\sin(\pi z) - \dot{\tau}_2\sin(2\pi z)]
\tag{5.20}
$$

where a dot now indicates the time derivative. Take the remaining terms to the right hand side

$$
P\frac{\partial \delta T}{\partial t} = \nabla^2\delta T + v_z - P\mathbf{v}\cdot\nabla\delta T
\tag{5.21}
$$

This expression can be evaluated with the aid of the following relations

$$
\begin{aligned}
\nabla^2\delta T &= -(\pi^2 + q^2)\tau_1\cos(qx)\sin(\pi z) + 4\pi^2\tau_2\sin(2\pi z) \\
v_z &= u\cos(qx)\sin(\pi z) \\
-Pv_x\frac{\partial \delta T}{\partial x} &= -P\pi u\tau_1\sin^2(qx)\cos(\pi z)\sin(\pi z) \\
-Pv_z\frac{\partial \delta T}{\partial z} &= -P\pi u\tau_1\cos^2(qx)\cos(\pi z)\sin(\pi z) \\
&\quad + P\,2\pi u\tau_2\cos(qx)\sin(\pi z)\cos(2\pi z)
\end{aligned}
\tag{5.22}
$$

In combining the last two terms to make $-P\mathbf{v}\cdot\nabla\delta T$, the term proportional to $u\tau_1$ is independent of x because $\cos^2(qx) + \sin^2(qx) = 1$ and the product $\cos(\pi z)\sin(\pi z)$ is simply the second harmonic $\frac{1}{2}\sin(2\pi z)$, which matches the assumed form in the second term of Eq. (5.9). As is typical of a quadratic nonlinearity like $\mathbf{v}\cdot\nabla\delta T$, the product of two spatial harmonics generates new sum and difference harmonics. The following trigonometric identity

$$
\sin(\pi z)\cos(2\pi z) = \frac{1}{2}[\sin(3\pi z) - \sin(\pi z)]
\tag{5.23}
$$

and some algebra then gives

$$
\begin{aligned}
-P\mathbf{v}\cdot\nabla\delta T &= -P\frac{\pi}{2}u\tau_1\sin(2\pi z) - P\pi u\tau_2\cos(qx)\sin(\pi z) \\
&\quad + P\pi u\tau_2\cos(qx)\sin(3\pi z)
\end{aligned}
\tag{5.24}
$$

41

The last nonlinear term on the right hand side of Eq. (5.24) contains $\sin(3\pi z)$, which represents a coupling to aOurier mode outside the set considered in Eqs. (5.19). We shall simply ignore the coupling to this higher spatial harmonic.[21] In the next section, we examine a model physical problem where the lower harmonics do indeed decouple from the rest. One can now equate the result in Eq. (5.20) with the right hand side given by Eqs. (5.21), (5.22) and (5.24). With the neglect of the last term in Eq. (5.24), the resulting expression contains only $\cos(qx)\sin(\pi z)$ and $\sin(2\pi z)$. Since these functions are linearly independent, one can simply equate their coefficients and find

$$
\begin{aligned}
P\dot{\tau}_1 &= -(\pi^2 + q^2)\tau_1 + u - P\pi u\tau_2 \\
-P\dot{\tau}_2 &= 4\pi^2\tau_2 - P\frac{\pi}{2}u\tau_1
\end{aligned}
\tag{5.25}
$$

This gives two coupled, nonlinear, ordinary differential equations in t for the three Fourier coefficients (τ_1, τ_2, u).

A third relation is obtained from the first of Eqs. (5.2). Make use of the following vector identity from Eq. (48.14)

$$
-(\mathbf{v}\cdot\nabla)\mathbf{v} = -\nabla\frac{1}{2}v^2 + \mathbf{v}\times(\nabla\times\mathbf{v})
\tag{5.26}
$$

to rewrite that equation as

$$
\frac{\partial\mathbf{v}}{\partial t} = -\nabla\left(\delta w + \frac{1}{2}v^2\right) + \nabla^2\mathbf{v} + R\hat{\mathbf{z}}\,\delta T + \mathbf{v}\times\boldsymbol{\zeta}
\tag{5.27}
$$

Here the vorticity $\boldsymbol{\zeta} = \nabla\times\mathbf{v} = \zeta\hat{\mathbf{y}}$ has again been introduced. With our truncated set of basis functions, Eq. (4.53) yields

$$
\zeta = \frac{u}{q}(\pi^2 + q^2)\sin(qx)\sin(\pi z)
\tag{5.28}
$$

The curl of Eq. (5.27) eliminates the gradient term (and the pressure!), leading to

$$
\frac{\partial\boldsymbol{\zeta}}{\partial t} = \nabla^2\boldsymbol{\zeta} + R\nabla\times(\hat{\mathbf{z}}\,\delta T) + \nabla\times(\mathbf{v}\times\boldsymbol{\zeta})
\tag{5.29}
$$

Some vector algebra establishes the following relations

$$
\begin{aligned}
\nabla\times(\hat{\mathbf{z}}\,\delta T) &= -\frac{\partial\delta T}{\partial x}\hat{\mathbf{y}} \\
\nabla\times(\mathbf{v}\times\boldsymbol{\zeta}) &= (\boldsymbol{\zeta}\cdot\nabla)\mathbf{v} - (\mathbf{v}\cdot\nabla)\boldsymbol{\zeta}
\end{aligned}
\tag{5.30}
$$

[21] An alternative, more systematic, procedure is to multiply the full equation [containing the term proportional to $\sin(3\pi z)$] separately by each of these two spatial harmonic functions $\cos(qx)\sin(\pi z)$ and $\sin(2\pi z)$ and then average over both x and z. In essence, this procedure, often known as the "Galerkin" method [Be78], projects the full equation onto the desired set of basis functions. In the present example, it eliminates the unwanted third harmonic and yields the desired coupled nonlinear equations.

where we use the relations $\nabla \cdot \mathbf{v} = 0$ and $\nabla \cdot \boldsymbol{\zeta} = 0$. Equation (5.29) can thus be rewritten

$$\frac{\partial \boldsymbol{\zeta}}{\partial t} = \nabla^2 \boldsymbol{\zeta} - R\frac{\partial \delta T}{\partial x}\hat{\mathbf{y}} + (\boldsymbol{\zeta} \cdot \nabla)\mathbf{v} - (\mathbf{v} \cdot \nabla)\boldsymbol{\zeta} \tag{5.31}$$

The y component of this relation then gives[22]

$$\frac{\partial \zeta}{\partial t} = \nabla^2 \zeta - R\frac{\partial \delta T}{\partial x} - (\mathbf{v} \cdot \nabla)\zeta \tag{5.32}$$

Substitution of Eq. (5.28) in this relation yields for the time derivative on the left hand side

$$\frac{\partial \zeta}{\partial t} = \frac{\pi^2 + q^2}{q}\dot{u}\sin(qx)\sin(\pi z) \tag{5.33}$$

The right hand side (r.h.s.) of Eq. (5.32) is

$$\text{r.h.s.} = \nabla^2 \zeta - R\frac{\partial \delta T}{\partial x} - (\mathbf{v} \cdot \nabla)\zeta \tag{5.34}$$

It is analyzed with the use of the following relations

$$\nabla^2 \zeta = -\frac{(\pi^2 + q^2)^2}{q}u\sin(qx)\sin(\pi z)$$
$$-R\frac{\partial \delta T}{\partial x} = R\tau_1 q\sin(qx)\sin(\pi z)$$
$$(\mathbf{v} \cdot \nabla)\zeta = 0 \tag{5.35}$$

where the last one follows from an explicit computation. Thus equating the time derivative in Eq. (5.33) with the right hand side, one finds

$$\dot{u} = -(\pi^2 + q^2)u + \frac{Rq^2}{(\pi^2 + q^2)}\tau_1 \tag{5.36}$$

Within this truncated set, the Navier-Stokes equation leads to a linear relation between (u, τ_1).

In *summary*, we have derived the following three coupled, nonlinear, first-order differential equations in t for the coefficients $[u(t), \tau_1(t), \tau_2(t)]$ of the truncated set of Fourier amplitudes in Eq. (5.19) in the Boussinesq approximation to the Rayleigh-Bénard problem in Fig. 3.1

$$\dot{u} = -(\pi^2 + q^2)u + \frac{Rq^2}{(\pi^2 + q^2)}\tau_1$$
$$P\dot{\tau}_1 = -(\pi^2 + q^2)\tau_1 + u - P\pi u \tau_2$$
$$-P\dot{\tau}_2 = 4\pi^2 \tau_2 - P\frac{\pi}{2}u \tau_1 \tag{5.37}$$

---

[22]Note that $(\boldsymbol{\zeta} \cdot \nabla)\mathbf{v} = (\zeta \, \partial/\partial y)\mathbf{v}$ vanishes here.

The nonlinear terms involving $(u\tau_1, u\tau_2)$ in the last two equations arise from convective heat transport. Apart from the basic truncation in Eq. (5.19), the only approximation here is the neglect of the coupling to the higher spatial harmonic $\sin(3\pi z)$ in the last of Eqs. (5.24), to which we will return in the next section.

It is useful to simplify the form of these equations by a change of variables. Introduce a rescaled time s and the ratio r of R to the lowest Rayleigh number $R_1(q)$ of the previous linearized problem

$$t \equiv \frac{Ps}{\pi^2 + q^2} \qquad ; R \equiv r\frac{(\pi^2 + q^2)^3}{q^2} = rR_1(q) \qquad (5.38)$$

Thus

$$\frac{du}{ds} = -Pu + Pr(\pi^2 + q^2)\tau_1$$

$$\frac{d\tau_1}{ds} = -\tau_1 + \frac{1}{\pi^2 + q^2}u - \frac{P\pi}{\pi^2 + q^2}u\,\tau_2$$

$$\frac{d\tau_2}{ds} = -\frac{4\pi^2}{\pi^2 + q^2}\tau_2 + \frac{P\pi}{2(\pi^2 + q^2)}u\,\tau_1 \qquad (5.39)$$

Then define

$$u \equiv \frac{\sqrt{2}\,(\pi^2 + q^2)}{P\pi}x \qquad ; \tau_1 \equiv \frac{\sqrt{2}}{P\pi r}y \qquad ; \tau_2 \equiv \frac{1}{P\pi r}z \qquad (5.40)$$

The result is

$$\dot{x} = -Px + Py$$

$$\dot{y} = -y + rx - xz$$

$$\dot{z} = -bz + xy \qquad (5.41)$$

The dot now indicates a derivative with respect to the rescaled time s. In these equations

$$b = \frac{4\pi^2}{\pi^2 + q^2} \qquad ; r = \frac{R}{R_1(q)} \qquad (5.42)$$

If we identify $q^2 = q_c^2$, where q_c^2 describes the periodic convective roll structure formed at the onset of instability in the linearized analysis for free-free surfaces, then

$$q^2 = q_c^2 = \pi^2/2$$
$$R_1(q_c) = R_c = 27\pi^4/4$$
$$b = 8/3 \qquad (5.43)$$

Equations (5.41) are known as the *Lorenz equations*. They form three first-order, coupled, nonlinear differential equations in rescaled time [Eq. (5.38)] for the Fourier coefficients (x, y, z) defined in Eqs. (5.40) and (5.19). The coefficient x characterizes the velocity field and (y, z) describe the temperature. There are three parameters in these equations (b, P, r). As seen above, b is determined entirely by the spatial periodicity $\lambda = 2\pi/q$, and $b = 8/3$ for $q = q_c$. The Prandtl number P is defined in Eq. (3.23) and depends on the ratio of the kinematic viscosity to the thermal diffusivity, two intrinsic properties of the fluid. The only remaining parameter in the Lorenz equations is then $r = R/R_1(q)$, where the Rayleigh number is defined in Eq. (3.23) and is determined by, among other things, the experimental parameter $h^3 \Delta T$ (Fig. 3.1). It is clear that r may be varied, for example, by varying the temperature difference ΔT between the bounding surfaces. The quantity r evidently functions as the control parameter in these equations. We examine the solutions to the Lorenz equations as a function of r in some detail in the next part of this book. These apparently simple equations are, in fact, quite remarkable.

6 Lorenz equations: direct derivation for simple physical configuration

The Lorenz equations are of sufficient importance that they merit an alternative derivation, which also isolates the essential elements and provides additional physical insight. We consider a more specific, constrained configuration where, with some additional simplifying assumptions, the dynamical equations take a relatively tractable form. The mathematical formulation in terms of Fourier amplitudes and Fourier coefficients then follows quite directly [Yo85]. This configuration has the added advantage that the time development of all the dynamical variables in the Lorenz equations can be clearly observed and monitored.

Let us go back to the Navier-Stokes and heat-flow Eqs. (60.27) and (60.48)

$$\rho \frac{d\mathbf{v}}{dt} = \rho \left[\frac{\partial \mathbf{v}}{\partial t} + (\mathbf{v} \cdot \boldsymbol{\nabla})\mathbf{v} \right] = -\boldsymbol{\nabla} p + \rho \mathbf{g} + \eta \nabla^2 \mathbf{v}$$
$$\frac{dT}{dt} = \frac{\partial T}{\partial t} + \mathbf{v} \cdot \boldsymbol{\nabla} T = \kappa \nabla^2 T \tag{6.1}$$

where $\mathbf{g} = -g\hat{\mathbf{z}}$. These equations assume incompressible, viscous flow in a gravitational field and neglect viscous heating.

Consider the configuration of Fig. 6.1 where the fluid flows in the narrow annular channel between two stationary, horizontal, concentric cylinders, the channel having a mean radius l measured from the common center. Assume that the fluid is heated from below and that both the inner and outer walls are maintained at the same fixed temperature $T_W(\phi)$ where ϕ is the polar angle measured from the downward vertical. We study flow that is uniform along the axis of the cylinders (namely, independent of y).

The velocity field for an incompressible viscous fluid flowing along a channel of width d with rigid walls is derived in [Fe03] and shown in Fig. 61.1, which we sketch in Fig. 6.2. The analytic form is

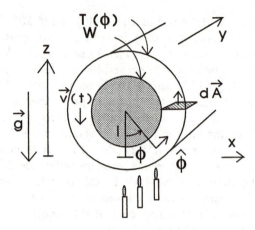

Fig. 6.1. Configuration for a direct derivation of the Lorenz equations. The fluid is free to move in the narrow region between two fixed, rigid, horizontal, concentric cylinders. The fluid is heated from below, and both walls are kept at a specified temperature $T_W(\phi)$. The fluid velocity is assumed to be uniform across the channel, and the effect of viscosity is approximated by a negative friction force proportional to that velocity. The temperature of the fluid is also assumed uniform across the channel, and the heat flow to and from the walls is taken to be proportional to the temperature difference between the fluid and the two walls. We study flow that is uniform along the axis of the cylinders (namely, independent of y).

$$\mathbf{v} = \frac{v_0}{(d/2)^2}[(d/2)^2 - x^2]\hat{\mathbf{z}}$$

$$\nabla^2\mathbf{v} = -2\frac{v_0}{(d/2)^2}\hat{\mathbf{z}} \tag{6.2}$$

The viscous term $\eta\nabla^2\mathbf{v}$ in the Navier-Stokes equation is negative and describes a friction force proportional to the velocity and opposing the flow. We simplify this spatially dependent velocity field to one that is uniform across the channel and independent of the azimuthal angle ϕ.

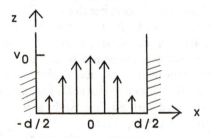

Fig. 6.2. Velocity field for steady viscous flow in a channel with rigid walls [after Fig. 61.1].

46

Thus, we assume the approximate form

$$\mathbf{v} = v(t)\hat{\boldsymbol{\phi}} \qquad (6.3)$$

which is obtained by a suitable average of the actual velocity field across the channel

$$\langle \eta \nabla^2 \mathbf{v} \rangle_{\text{channel}} = -\mathcal{R}\rho\mathbf{v} \qquad (6.4)$$

Here, $\mathcal{R}$ is a fluid constant with the dimension of an inverse time (on dimensional grounds, it is of order $\mathcal{R} \sim \nu/d^2$). This velocity field evidently satisfies $\nabla \cdot \mathbf{v} = \partial v(t)/l\partial\phi = 0$.

The temperature pattern for fluid flowing in a channel with a specified temperature at the walls will have a spatial distribution across the channel analogous to that shown in Fig. 6.2. For simplicity, we also assume that the temperature in the fluid is *uniform across the channel*, and that the heat flow to and from the walls is proportional to the local temperature difference between the fluid and the walls

$$T = T(\phi, t)$$
$$\langle \kappa \nabla^2 T \rangle_{\text{channel}} = -K[T(\phi, t) - T_W(\phi)] \qquad (6.5)$$

where K is a fluid constant that has the dimension of an inverse time (on dimensional grounds, $K \sim \kappa/d^2$).

One special type of density change in the fluid must be included—that associated with thermal expansion. When the fluid is heated, its density decreases, and this alteration provides an additional buoyant force on the fluid at the bottom of the channel. In the Boussinesq approximation, only the linear temperature dependence is retained. Thus the gravitational force term in the Navier-Stokes equation will be approximated as

$$\rho\mathbf{g} \approx \rho_0\{1 - \beta[T(\phi, t) - T_0]\}\mathbf{g} \qquad (6.6)$$

Here β is the thermal expansion coefficient and T_0 is some reference temperature. Otherwise, incompressible flow is assumed with $\rho = \rho_0$. Under these simplifying approximations the Navier-Stokes and heat-flow equations for the physical configuration in Fig. 6.1 then take the form

$$\rho_0\left[\frac{\partial \mathbf{v}}{\partial t} + (\mathbf{v} \cdot \nabla)\mathbf{v}\right] = -\nabla p + \rho_0[1 - \beta(T - T_0)]\mathbf{g} - \mathcal{R}\rho_0\mathbf{v}$$
$$\frac{\partial T}{\partial t} + \mathbf{v} \cdot \nabla T = -K[T - T_W(\phi)] \qquad (6.7)$$

Note that the full convective nonlinearity is retained in both equations.

We now seek a solution to these equations of the form

$$\mathbf{v} = v(t)\hat{\boldsymbol{\phi}}$$
$$p = p(\phi, t)$$
$$T = T(\phi, t) \qquad (6.8)$$

47

Let $d\mathbf{A} = \hat{\phi}\,dA$ be an element of transverse area across the fluid channel (Fig. 6.1). Dot the following vector operator $(2\pi\rho_0)^{-1}\int_0^{2\pi} d\phi \int_A d\mathbf{A}$ into the first of Eqs. (6.7), and use the following results

$$\hat{\phi}\cdot\boldsymbol{\nabla} = \frac{1}{l}\frac{\partial}{\partial\phi}$$

$$\hat{\phi}\cdot\frac{\partial\hat{\phi}}{\partial\phi} = \frac{\partial}{\partial\phi}\left(\frac{1}{2}\hat{\phi}\cdot\hat{\phi}\right) = 0 \tag{6.9}$$

Then

$$\int_A d\mathbf{A}\cdot\mathbf{v} = A\,v(t) \equiv Q(t) \quad ; \text{ flow rate}$$

$$\int_A d\mathbf{A}\cdot(\mathbf{v}\cdot\boldsymbol{\nabla})\mathbf{v} = \int_A d\mathbf{A}\cdot\left\{v(t)\frac{1}{l}\frac{\partial[v(t)\hat{\phi}]}{\partial\phi}\right\} = 0$$

$$\frac{A}{2\pi\rho_0\,l}\int_0^{2\pi} d\phi\frac{\partial p(\phi,t)}{\partial\phi} = 0 \tag{6.10}$$

Here $Q(t)$ is an algebraic quantity that gives the flow rate of the fluid through the transverse area of the channel. Note that the convective term in the Navier-Stokes equation does not contribute for this geometry. Note further that the pressure also disappears from the problem since it is single-valued in ϕ! For the gravitational term use

$$\mathbf{g}\cdot d\mathbf{A} = -gdA\sin\phi$$

$$\int_0^{2\pi}\sin\phi\,d\phi = 0 \tag{6.11}$$

For the second of Eqs. (6.7) use

$$(\mathbf{v}\cdot\boldsymbol{\nabla})T(\phi,t) = \frac{Q(t)}{Al}\frac{\partial T(\phi,t)}{\partial\phi} \tag{6.12}$$

With the use of these results, Eqs. (6.7) become

$$\frac{dQ(t)}{dt} = \frac{\beta gA}{2\pi}\int_0^{2\pi} T(\phi,t)\sin\phi\,d\phi - \mathcal{R}\,Q(t)$$

$$\frac{\partial T(\phi,t)}{\partial t} + \frac{Q(t)}{Al}\frac{\partial T(\phi,t)}{\partial\phi} = K[T_W(\phi) - T(\phi,t)] \tag{6.13}$$

The problem has now been reduced to two coupled, nonlinear, partial integro-differential equations for the fluid flow rate and temperature $[Q(t), T(\phi,t)]$. The nonlinearity enters through the convective part of the heat equation.

Assume that the temperature inversion in Fig. 6.1 has a linear z dependence and write[23]

$$T_W(\phi) \equiv T_0 + T_1\cos\phi \quad ; T_1 > 0 \tag{6.14}$$

[23]Note that in the thin channel $z \approx l(1-\cos\phi)$.

where the zero of temperature is arbitrary. This choice means that the temperature of the walls is right-left symmetric.

The physical configuration in Fig. 6.1 is now manifestly periodic in ϕ over the interval $[0, 2\pi]$, and one is thus led directly to a Fourier-series representation of the solution to the problem posed in Eqs. (6.13). Write

$$T(\phi, t) = T_0 + c_0(t) + \sum_{n=1}^{\infty} [s_n(t) \sin(n\phi) + c_n(t) \cos(n\phi)] \qquad (6.15)$$

Now substitute this expansion in the second of Eqs. (6.13). Since the functions $\sin(n\phi)$ and $\cos(n\phi)$ are linearly independent, one can simply equate their coefficients. For $n = 0$ one has the elementary solution

$$\frac{dc_0}{dt} = -Kc_0$$
$$c_0(t) = c_0(0)e^{-Kt} \qquad ; n = 0 \qquad (6.16)$$

Evidently, this contribution is a transient that dies off exponentially with a characteristic decay time K^{-1}.

For $n > 1$ one has the pair of equations

$$\frac{ds_n}{dt} - \frac{nQ}{Al}c_n = -Ks_n$$
$$\frac{dc_n}{dt} + \frac{nQ}{Al}s_n = -Kc_n \qquad ; n > 1 \qquad (6.17)$$

Multiply the first equation by s_n and the second by c_n and add

$$s_n\frac{ds_n}{dt} + c_n\frac{dc_n}{dt} = -K(s_n^2 + c_n^2) \qquad (6.18)$$

Now define the positive-definite combination of the square of these Fourier coefficients (the squared amplitude)

$$a_n^2 \equiv s_n^2 + c_n^2 \qquad (6.19)$$

Equation (6.18) then states that

$$\frac{da_n^2}{dt} = -2Ka_n^2$$
$$a_n^2(t) = a_n^2(0)e^{-2Kt} \qquad ; n > 1 \qquad (6.20)$$

This contribution is also a transient, and it implies that both (s_n, c_n) die off with time if $n > 1$.

When $Kt \gg 1$, we see that only the $n = 1$ contribution remains, and therefore the $n = 1$ contribution *decouples from the rest of the problem!*[24]

$$T(\phi, t) \rightarrow T_0 + s_1(t) \sin\phi + c_1(t) \cos\phi \qquad ; Kt \gg 1 \qquad (6.21)$$

---

[24]Note that this result arises from the specific form of the nonuniform temperature distribution of the walls $T_W = T_0 + T_1 \cos\phi$. As seen in Eq. (6.23), the control parameter r is proportional to T_1, which characterizes the magnitude of the temperature nonuniformity.

All the long-time physics then resides in the $n = 1$ contribution. When $Kt \gg 1$, Eqs. (6.13) thus become[25]

$$\frac{dQ(t)}{dt} = \frac{\beta g A}{2} s_1(t) - \mathcal{R}Q(t)$$

$$\frac{ds_1(t)}{dt} - \frac{Q(t)}{Al} c_1(t) = K[-s_1(t)]$$

$$\frac{dc_1(t)}{dt} + \frac{Q(t)}{Al} s_1(t) = K[T_1 - c_1(t)] \qquad\qquad ; n = 1 \qquad (6.22)$$

These have exactly the structure of the *Lorenz equations*. To recover the previous form, introduce the dimensionless parameters

$$r \equiv \frac{g\beta T_1}{2K\mathcal{R}l}$$

$$P \equiv \frac{\mathcal{R}}{K}$$

$$s = Kt \qquad\qquad (6.23)$$

The first two are the scaled effective Rayleigh number and the Prandtl numbers.[26] The last is the dimensionless rescaled time. Introduce also the dimensionless functions

$$x \equiv \frac{Q(t)}{KAl} \qquad\qquad ; \text{flow rate}$$

$$y \equiv \frac{rs_1(t)}{T_1} \qquad\qquad ; \text{temperature variation at } \phi = \pi/2$$

$$z \equiv \frac{r[T_1 - c_1(t)]}{T_1} \qquad\qquad ; \text{temperature variation at } \phi = 0 \qquad (6.24)$$

The first is the fluid flow rate in the channel. From Eq. (6.21), the quantities (s_1, c_1) in the second and third give the temperature variations at $\phi = \pi/2$ and $\phi = 0$, respectively.

After some algebra, Eqs. (6.22) then take the familiar form

$$\dot{x} = -Px + Py$$

$$\dot{y} = -y + rx - xz$$

$$\dot{z} = -bz + xy \qquad\qquad ; b = 1 \qquad (6.25)$$

Here the dot indicates a derivative with respect to the dimensionless time s. Apart from the geometrical parameter $b = 1$ that is specific to the present model, these coupled nonlinear equations are precisely the Lorenz Eqs. (5.41).

---

[25]Use $\int_0^{2\pi} \sin^2 \phi \, d\phi = \pi$ and $\int_0^{2\pi} \sin \phi \cos \phi \, d\phi = 0$.

[26]As noted from Eqs. (6.4) and (6.5), to within numerical constants, $\mathcal{R} \sim \nu/d^2$ and $K \sim \kappa/d^2$. With $2T_1 = \Delta T$, Eqs. (6.23) become $P \sim \nu/\kappa$ and $r \sim (g\beta\Delta T/\kappa\nu)(d^4/4l)$, which now compare directly with Eqs. (3.23).

Part III
Discrete Dynamical Systems

7 Example of a nonlinear oscillator

The first part of [Fe03] deals with particle mechanics, the study of dynamical systems with a finite number of degrees of freedom. We correspondingly turn here to the subject of the nonlinear dynamics of such systems. As groundwork, we study the Duffing oscillator, which is a one-dimensional simple harmonic oscillator with an additional quartic term in the potential.

Duffing oscillator: general form

The Duffing oscillator is a very common example of nonlinear dynamics in which the familiar harmonic quadratic potential is augmented by a quartic term[27]

$$V_D(q) = \frac{1}{2}m\alpha q^2 + \frac{1}{4}m\beta q^4 \tag{7.1}$$

This potential provides a one-dimensional model for the theory of second-order phase transitions [La80], introduced by Landau in the 1930s and widely used in the description of superconducting metals through the Ginzburg-Landau theory. It also reappears in the σ-model for spontaneously broken chiral symmetry in the strong interactions, and in the Higgs mechanism for generating the mass of the gauge bosons in the standard model of electroweak interactions (see, for example, [Wa04]). The parameter β is generally assumed to be positive, ensuring that $V_D(q)$ remains positive for large $|q|$. In contrast, the parameter α can in principle have either sign, and the potential V_D displays remarkable behavior as α is tuned continuously from a positive value through zero to a negative value. In the context of phase transitions, one thinks of $\alpha(T)$ as being temperature-dependent. Specifically, $\alpha(T)$ vanishes linearly at some critical transition temperature T_c. Thus $\alpha(T_c) = 0$, and $\alpha(T)$ has the same algebraic sign as $T - T_c$.

It is helpful to seek the stationary points of V_D defined by the equation

$$\frac{dV_D(q)}{dq} = m\left(\alpha q + \beta q^3\right) = 0 \tag{7.2}$$

This equation has three roots given by

$$q = 0 \qquad ; q^2 = -\frac{\alpha}{\beta} \tag{7.3}$$

The first root at the origin represents an allowed physical position for all values of α. In contrast, the other roots at $q = \pm\sqrt{-\alpha/\beta}$ are imaginary if $\alpha > 0$ (assuming $\beta > 0$), but they become physical positions $\pm\sqrt{|\alpha|/\beta}$ if α is negative.

We therefore consider the following distinct possibilities with positive β (see Fig. 7.1)

---

[27]We assume for the purposes of this general discussion that m has the dimensions of energy $[ml^2/t^2]$ and that (α, β, q) are dimensionless.

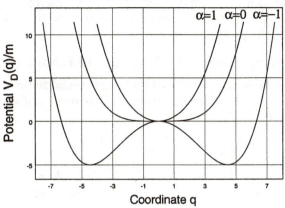

DUFFING POTENTIAL

Fig. 7.1. Duffing Oscillator potential $V_D(q)/m = \frac{1}{2}\alpha q^2 + \frac{1}{4}\beta q^4$ for $\beta = 0.05$ and three values of $\alpha = (+1, 0, -1)$.

1. $\alpha > 0$:

   In this case, V_D has only one stationary point at $q = 0$, since both the quadratic and quartic contributions increase with increasing q^2. Thus the origin is the absolute minimum of $V_D(q)$.

2. $\alpha < 0$:

   In this case, there are three stationary points, at $q = 0$ and $\pm\sqrt{|\alpha|/\beta}$. Since the quadratic contribution now decreases with increasing q^2, the potential $V_D(q)$ decreases symmetrically from the origin, which is a local maximum. $V_D(q)$ reaches symmetric absolute minima at $q = \pm\sqrt{|\alpha|/\beta}$, beyond which the quartic contribution starts to dominate, and $V_D(q)$ eventually becomes positive for large q^2.

3. $\alpha = 0$:

   In this case, the quadratic potential term is absent, so that $V_D(q)$ is very flat around the origin, rising rapidly only when βq^4 is of order one. For this special value of α, the potential provides very small restoring force for small $|q|$, and the oscillation frequency for periodic motion would be correspondingly small.

We see that the qualitative character of the Duffing potential $V_D(q)$ depends critically on the sign of the quadratic contribution. Such behavior exemplifies a common phenomenon in nonlinear physics. For positive α, there is a single absolute minimum, but it "bifurcates" at $\alpha = 0$ into one of two symmetric absolute minima for negative α. Thus, the equilibrium position of the system shifts continuously from the origin to either of the two minima as soon as α changes sign from positive to negative. Without additional physical assumptions, there is no reason to prefer one or the other of the two equivalent positions (this situation is known as a "broken symmetry"), but there is an energy barrier of height $\Delta V_D = \frac{1}{4}m\alpha^2/\beta$ between them. Some other aspects of the Duffing potential are investigated in the problems.

Duffing oscillator: perturbed simple harmonic oscillator

We now return to usual dimensional units and specialize the general Duffing potential to consider a one-dimensional simple harmonic oscillator with a small additional quartic term (Fig. 7.2)

$$V(q) = \frac{1}{2}m\omega_0^2 q^2 + \frac{1}{4}\varepsilon m q^4 \tag{7.4}$$

We assume here that ε is small, although we will allow either sign.[28] The corresponding dynamical equation of motion and hamiltonian are

$$\ddot{q} + \omega_0^2 q + \varepsilon q^3 = 0$$

$$H = \frac{p^2}{2m} + \frac{1}{2}m\omega_0^2 q^2 + \frac{1}{4}\varepsilon m q^4 \tag{7.5}$$

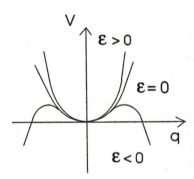

Fig. 7.2. Sketch of potential of the perturbed simple harmonic oscillator for various values of ε.

The most straightforward approach to finding an analytic solution is to assume a power series in ε and work consistently to a given order. Thus we write

$$q(t) = q_0(t) + \varepsilon q_1(t) + \cdots \tag{7.6}$$

Substitution in the first of Eqs. (7.5) and identification of coefficients of powers of ε yield the following equations through first order

$$\ddot{q}_0 + \omega_0^2 q_0 = 0 \qquad ; \text{ zero-order}$$

$$\ddot{q}_1 + \omega_0^2 q_1 + q_0^3 = 0 \qquad ; \text{ first-order} \tag{7.7}$$

---

[28]We focus on the small change in the dynamics for motion near the origin. In this case, the motion remains bounded, so that the negative (and hence unstable) potential for negative ε and large q^2 is irrelevant here; note that the parameter ε now has dimensions $[1/l^2 t^2]$.

Assume a set of initial conditions

$$q(0) = a$$
$$\dot{q}(0) = 0 \qquad (7.8)$$

so that the oscillator starts from rest. The solution to the zero-order simple harmonic oscillator equation is

$$q_0(t) = a \cos \omega_0 t$$
$$q_0(0) = a \qquad ; \; \dot{q}_0(0) = 0 \qquad (7.9)$$

The initial conditions on the remaining contributions to $q(t)$ are

$$q_i(0) = 0$$
$$\dot{q}_i(0) = 0 \qquad ; \; i = 1, 2, \cdots \qquad (7.10)$$

Substitution of the zero-order result into the second of Eqs. (7.7) gives

$$\ddot{q}_1 + \omega_0^2 q_1 = -a^3 \cos^3 \omega_0 t \qquad (7.11)$$

A little algebra establishes the following trigonometric identity

$$\cos^3 \omega_0 t = \frac{1}{4} \left(\cos 3\omega_0 t + 3 \cos \omega_0 t \right) \qquad (7.12)$$

Thus Eq. (7.11) can be rewritten as

$$\ddot{q}_1 + \omega_0^2 q_1 = -\frac{3a^3}{4} \cos \omega_0 t - \frac{a^3}{4} \cos 3\omega_0 t \qquad (7.13)$$

This equation describes a harmonically driven oscillator, but note that there are driving terms at frequencies of both ω_0 and $3\omega_0$. The nonresonant term with $\cos 3\omega_0 t$ is easy to handle. In contrast, as we shall see, the resonant contribution proportional to $\cos \omega_0 t$ leads to *secular* terms in the coordinate $q(t)$ that grow without bound.

It is readily verified that the solution to Eq. (7.13) satisfying the initial conditions in Eqs. (7.10) is

$$q_1(t) = -\frac{a^3}{8\omega_0^2} \left[3\omega_0 t \sin \omega_0 t + \frac{1}{4} \left(\cos \omega_0 t - \cos 3\omega_0 t \right) \right] \qquad (7.14)$$

To order ε, the total solution for the coordinate $q(t)$ that satisfies the initial conditions in Eqs. (7.8) is

$$q(t) = a \cos \omega_0 t - \frac{\varepsilon a^3}{8\omega_0^2} \left[3\omega_0 t \sin \omega_0 t + \frac{1}{4} \left(\cos \omega_0 t - \cos 3\omega_0 t \right) \right] + O(\varepsilon^2) \qquad (7.15)$$

This result cannot be the correct solution for large t, however, since we know that the actual solution must remain bounded. This conclusion follows from energy conservation[29]

$$E = \frac{1}{2} m \dot{q}^2 + \frac{1}{2} m \omega_0^2 q^2 + \frac{1}{4} \varepsilon m q^4 = \text{constant} \qquad (7.16)$$

[29] If $\varepsilon < 0$, the coordinate is bounded provided we are below the barrier (see Fig. 7.2).

Evidently, straightforward perturbation theory fails when $\omega_0 t \approx 1$. What is wrong? The resonant driving term above has the same resonant frequency ω_0. What we forgot is that the nonlinear oscillator also has a *frequency shift* $\omega_0 \to \omega$. Let us therefore expand

$$\omega(\varepsilon) = \omega_0 + \varepsilon \omega_1 + \cdots \qquad (7.17)$$

so that $q(t)$ has the approximate form $a(\varepsilon) \cos [\omega(\varepsilon)t]$. To $O(\varepsilon)$, this time dependence must also be expanded

$$
\begin{aligned}
q(t) &= a(\varepsilon) \cos [\omega(\varepsilon)t] \\
&= a(\varepsilon) \cos [\omega_0 t + \varepsilon \omega_1 t + \cdots] \\
&= a(\varepsilon) [\cos \omega_0 t - \varepsilon \omega_1 t \sin \omega_0 t + \cdots] \qquad (7.18)
\end{aligned}
$$

which accounts for the presence of the secular term proportional to $t \sin \omega_0 t$.

We shall now repeat the previous analysis employing the expansions in Eqs. (7.6) and (7.17). We assume that the solution only depends on time through the combination $\tau = \omega t$ and write the full equation of motion in Eq. (7.5) as

$$\omega^2 \frac{d^2 q}{d\tau^2} + \omega_0^2 q + \varepsilon q^3 = 0 \qquad ; \tau = \omega t \qquad (7.19)$$

To $O(\varepsilon)$, substitution of the expansion in Eq. (7.17) into this equation then gives

$$\omega_0^2 \frac{d^2 q}{d\tau^2} + \omega_0^2 q + 2\varepsilon \omega_0 \omega_1 \frac{d^2 q}{d\tau^2} + \varepsilon q^3 + \cdots = 0 \qquad (7.20)$$

The expansion in Eq. (7.6) then allows us to identify the new zero-order and first-order equations

$$
\begin{aligned}
\omega_0^2 \left(\frac{d^2 q_0}{d\tau^2} + q_0 \right) &= 0 \\
\omega_0^2 \left(\frac{d^2 q_1}{d\tau^2} + q_1 \right) &= -2\omega_0 \omega_1 \frac{d^2 q_0}{d\tau^2} - q_0^3 \qquad (7.21)
\end{aligned}
$$

The first equation is the same simple harmonic oscillator. The solution that satisfies the initial conditions can be inserted in the second equation to give the equations [compare Eqs. (7.9) and (7.13)]

$$
\begin{aligned}
q_0(t) &= a \cos \tau \\
\omega_0^2 \left(\frac{d^2 q_1}{d\tau^2} + q_1 \right) &= \left(2\omega_0 \omega_1 - \frac{3}{4}a^2 \right) a \cos \tau - \frac{1}{4}a^3 \cos 3\tau \qquad (7.22)
\end{aligned}
$$

As before, the resonant driving term proportional to $\cos \tau$ in the second equation would lead to secular growth in the coordinate, but we now have the possibility of *eliminating* it by a judicious choice in the frequency shift. Take

$$\omega_1 = \frac{3a^2}{8\omega_0} \qquad (7.23)$$

We can now set $\tau = \omega_0 t$ on the r.h.s. of the second equation since we are dealing with small corrections. The solution to the remaining equation then follows as in Eq. (7.14)

$$q_1(t) = -\frac{a^3}{32\omega_0^2}(\cos\omega_0 t - \cos 3\omega_0 t) \tag{7.24}$$

Hence the full solution through $O(\varepsilon)$ is found to be

$$q(t) = a\cos\left[\left(\omega_0 + \varepsilon\frac{3a^2}{8\omega_0}\right)t\right] - \frac{\varepsilon a^3}{32\omega_0^2}(\cos\omega_0 t - \cos 3\omega_0 t) + \cdots \tag{7.25}$$

The approximate anharmonic frequency is given by

$$\omega = \omega_0 + \varepsilon\frac{3a^2}{8\omega_0} \tag{7.26}$$

It depends on both ε and the amplitude a^2. If $\varepsilon > 0$, the quartic term raises the potential and the frequency increases; if $\varepsilon < 0$, the quartic term reduces the potential and the frequency decreases (see Fig. 7.2).

If the first term in Eq. (7.25) is expanded in powers of ε (which we do *not* do), the previous secular term in Eq. (7.15) is recovered. By keeping this term inside the argument of the cos, one has selectively summed all orders in perturbation theory to provide a solution for $q(t)$ that holds for times $\omega_0 t \gg 1$.

To understand the validity of this analysis, one can compare the analytic expression in Eq. (7.25) with the result obtained by direct numerical integration of the nonlinear differential equation. Introduce the dimensionless variables

$$x = \omega_0 t \qquad ; y = \frac{q}{a} \qquad ; \lambda = \frac{\varepsilon a^2}{\omega_0^2} \tag{7.27}$$

The first of Eqs. (7.5) and Eqs. (7.8) then become

$$\frac{d^2y}{dx^2} + y + \lambda y^3 = 0$$
$$y(0) = 1 \qquad ; y'(0) = 0 \tag{7.28}$$

In this form, numerical integration is readily carried out.[30]

The analytic expression in Eq. (7.25) and perturbation theory result in Eq. (7.15) take the form

$$y = \cos\left[x\left(1 + \frac{3\lambda}{8}\right)\right] - \frac{\lambda}{32}(\cos x - \cos 3x) \qquad ; \text{analytic}$$

$$y = \cos x - \frac{\lambda}{8}\left[3x\sin x + \frac{1}{4}(\cos x - \cos 3x)\right] \qquad ; \text{pert. theory} \tag{7.29}$$

---

[30]The numerical results presented here, and subsequent ones, were obtained with Mathcad11 using the Runge-Kutta algorithm.

The results are compared for a representative value of $\lambda = 0.6$ in Fig. 7.3. Note how well the analytic expression tracks the numerical result, even with a relatively large dimensionless coupling constant $\lambda = 0.6$ for the quartic term, and how quickly the perturbation theory expression deviates from it. A small, next-order correction to the angular frequency could make the analytic expression indistinguishable from the numerical one over this range.[31]

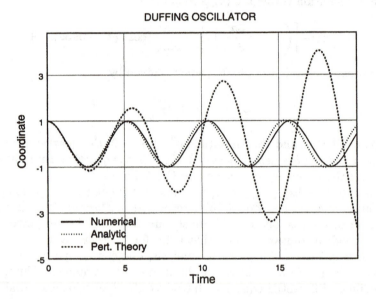

Fig. 7.3. Duffing Oscillator. The result from numerical integration of Eqs. (7.28) is compared with the analytic and perturbation theory expressions in Eqs. (7.29). The ordinate is q/a and the abscissa $\omega_0 t$. The curves are for $\lambda = \varepsilon a^2/\omega_0^2 = 0.6$.

The discussion in this section illustrates the crucial role played by resonance in nonlinear dynamics.

8 Phase-space dynamics and fixed points

We start this discussion with some familiar one-dimensional systems. Recall the concept of *phase space*. Given a system with hamiltonian $H(p, q, t)$, Hamilton's equations are

$$\frac{dp}{dt} = -\frac{\partial H}{\partial q}$$

$$\frac{dq}{dt} = \frac{\partial H}{\partial p} \qquad ; \frac{dH}{dt} = \frac{\partial H}{\partial t} \qquad (8.1)$$

Here the partial derivative implies that all other variables in H are kept constant.

---

[31] Improved approximation schemes are presented in [Be78a, Am03].

For definiteness, consider a time-independent hamiltonian $H(p, q)$, so that H is a constant of the motion. The instantaneous configuration of the system is completely specified by a point $[p(t), q(t)]$ in the (here) two-dimensional phase space. Given an initial point $[p(0), q(0)] \equiv (p_0, q_0)$, Hamilton's equations then provide a set of two coupled, first-order differential equations in the time that map out a trajectory in this phase space. In general, that trajectory is unique, which remains true near an unstable fixed point, even though the trajectory can proceed in different directions along the actual separatrices (as seen in Fig. 1.1). Two familiar examples are:

1. The one-dimensional simple harmonic oscillator

$$
\begin{aligned}
H(p, q) &= \frac{p^2}{2m} + \frac{1}{2}m\omega_0^2 q^2 \\
\dot{p} &= -m\omega_0^2 q \\
m\dot{q} &= p
\end{aligned}
\tag{8.2}
$$

2. The planar pendulum of length l

$$
\begin{aligned}
H(p_\theta, \theta) &= \frac{p_\theta^2}{2ml^2} + mgl(1 - \cos\theta) \\
\dot{p}_\theta &= -mgl\sin\theta \\
ml^2\dot{\theta} &= p_\theta
\end{aligned}
\tag{8.3}
$$

Here θ is the angle measured from the vertical, down position.

The first example provides coupled, *linear* equations of motion, and the second example provides coupled *nonlinear* equations of motion. Since the hamiltonian is the energy in both cases, motion along a particular trajectory satisfies

$$
\begin{aligned}
\frac{p^2}{2m} + \frac{1}{2}m\omega_0^2 q^2 &= E = \text{constant} && \text{; oscillator} \\
\frac{p_\theta^2}{2ml^2} + mgl(1 - \cos\theta) &= E = \text{constant} && \text{; pendulum} \\
&\xrightarrow{\theta \to 0} \frac{p_\theta^2}{2ml^2} + \frac{mgl}{2}\theta^2
\end{aligned}
\tag{8.4}
$$

In the first case, the phase-space orbit is an ellipse for all initial conditions, and it is easy to see from the equations of motion that the phase point follows the orbit in a clockwise direction. For the pendulum with small θ, the phase-space orbit is also an ellipse, because the problem then reduces to the simple harmonic oscillator.

Suppose, however, that the pendulum has the following configuration ($\theta = \pm\pi, p_\theta = 0$), namely pointing straight up and at rest. This configuration is clearly a *stationary*, or *fixed point*, since the time derivatives of both coordinates then vanish

$$
\begin{aligned}
\dot{p}_\theta &= 0 \\
\dot{\theta} &= 0 && \text{; } \theta = \pm\pi
\end{aligned}
\tag{8.5}
$$

59

It is also evident that there are two possible trajectories, since the pendulum can then either fall forward or backward.

To investigate these possibilities, let us expand to first order about the stationary point

$$p_\theta = \delta p_\theta$$
$$\theta = \pm\pi + \delta\theta \tag{8.6}$$

In this case, the linearized equations of motion become

$$\delta\dot{p}_\theta = mgl\,\delta\theta$$
$$ml^2\,\delta\dot{\theta} = \delta p_\theta \tag{8.7}$$

We seek a solution to these equations of the form (we anticipate a real γ)

$$\delta p_\theta = A\,e^{\gamma t} \quad ; \quad \delta\theta = B\,e^{\gamma t} \tag{8.8}$$

The linearized equations then reduce to

$$\gamma A - mgl\,B = 0$$
$$-A + ml^2\gamma\,B = 0 \tag{8.10}$$

These linear homogeneous algebraic equations will have a nontrivial solution for the amplitudes (A, B) only if the determinant of the coefficients vanishes

$$ml^2\gamma^2 - mgl = 0$$
$$\gamma = \pm\sqrt{\frac{g}{l}} = \pm\omega_0 \tag{8.11}$$

where ω_0 is the angular frequency of the small-amplitude motion for small displacements around the equilibrium down position (here, ω_0 simply provides an inverse time for dimensional purposes). Substitution back into either of Eqs. (8.10) then gives the ratio of the amplitudes for these two values of γ

$$\frac{A}{B} = \pm m\sqrt{gl^3} = \pm ml^2\omega_0 \tag{8.12}$$

Each of the trajectories through the fixed point defines a *separatrix*. If the plus sign is chosen in these last two equations, the trajectory moves *away* from the fixed point along a line with slope $ml^2\omega_0$, indicating instability (this result applies to both sides of the line). With the minus sign, the orbit moves *toward* the fixed point along a line with slope $-ml^2\omega_0$ (again, this behavior holds for both sides of the line). For both branches of the separatrix, the energy is $E = 2mgl$, which is the difference in potential energy from the bottom to the top of the pendulum in Eqs. (8.4). These features are evident in Fig. 1.1. The behavior of this linearized dynamical system is studied in more detail in the problems.

If the energy E is smaller than $2mgl$, then the pendulum will oscillate around the downward position ($\theta = 0$), and the angle θ remains bounded. If the energy E is larger than $2mgl$, then the pendulum goes around in a circle ("over the top"), and the angle increases continously.

These examples have identified two types of *periodic motion*. When the coordinates (p, q) return to the same set of initial conditions (because of energy conservation), the trajectory necessarily repeats itself.[32] The two types are:

1. *Libration.* Here the phase-space orbit is closed and the trajectory cycles around that orbit. The coordinate θ oscillates, continously retracing its path. Examples are the two orbits in Eqs. (8.4), provided that the energy of the pendulum satisfies $E < 2mgl$;

2. *Rotation.* Here the phase-space orbit is open and the trajectory repeats itself throughout phase space. The coordinate θ increases (or decreases) continously (note that the hamiltonian and the physical configuration are periodic in θ with period 2π, so that $\theta + 2\pi n$ is completely equivalent to θ for any integer n). An example is the pendulum orbit for $E > 2mgl$.

In these examples, the *separatrix* is an orbit through the unstable fixed point that divides phase space between these two types of motion (for a pendulum, there are two such orbits). All these features of the phase space of the simple, planar pendulum were previously illustrated in Fig. 1.1. A more general discussion of fixed points of nonlinear dynamical systems is given at the end of this section, including the stability of motion near a fixed point. These ideas are important in the subsequent analysis of the Lorenz equations and of other nonlinear equations.

Action-angle variables

Action-angle variables provide an alternative dynamical description of periodic systems. They play a central role in subsequent developments, and we here review them within the context of these simple examples. Consider a conservative system with

$$H(p, q) \;=\; \frac{p^2}{2m} + V(q) \;=\; E \qquad ; \text{ constant} \qquad (8.13)$$

The corresponding Hamilton-Jacobi Eq. (35.12) is

$$\frac{1}{2m}\left(\frac{\partial S}{\partial q}\right)^2 + V(q) + \frac{\partial S}{\partial t} \;=\; 0 \qquad (8.14)$$

A solution of this partial differential equation provides a function $S(q, P, t)$ that generates a canonical transformation to a new set of coordinates (P, Q), both of which are constants of the motion [Eqs. (35.10) and (35.11)]

$$P \;=\; \text{constant}$$
$$Q \;=\; \frac{\partial S(q, P, t)}{\partial P} \;=\; \text{constant} \qquad (8.15)$$

---

[32]The *period* of the motion is the time that it takes to complete one cycle.

The solution $S(q, P, t)$ also gives

$$p = \frac{\partial S(q, P, t)}{\partial q} \tag{8.16}$$

In the present case, one can solve the Hamilton-Jacobi (H-J) equation by assuming $P = E$ and taking

$$S(q, E, t) = W(q, E) - Et \tag{8.17}$$

where $W(q, E)$ is referred to as Hamilton's characteristic function. Equation (35.26) then follows

$$\frac{\partial S}{\partial t} = -E$$

$$\frac{1}{2m}\left(\frac{dW}{dq}\right)^2 + V(q) = E \tag{8.18}$$

Suppose that the motion is *periodic*, as in the two examples above. Define the *action* as[33]

$$J = \frac{1}{2\pi}\oint_{\text{cycle}} p\, dq \tag{8.19}$$

Here the integral is over one complete cycle; it is an area in phase space. Equations (8.16) and (8.17) combine to give

$$p = \frac{\partial W(q, E)}{\partial q}$$

$$J = \frac{1}{2\pi}\oint_{\text{cycle}} \frac{\partial W(q, E)}{\partial q}\, dq = J(E) \tag{8.20}$$

Evidently, the q dependence has now been integrated out, leaving E as the only variable in $J(E)$. One is free to choose *any* constant for P in H-J theory. In particular, choose $J(E)$ instead of E, and assume that this function can be inverted to give $E(J)$ (we shall show below how this is done). Write $W = W[q, E(J)]$, taking J as the new constant. Call this function $\bar{W}(q, J)$. Consequently

$$
\begin{aligned}
S[q, E(J), t] &= \bar{W}(q, J) - E(J)t \\
&\equiv \bar{S}(q, J, t)
\end{aligned} \tag{8.21}
$$

Now use this function $\bar{S}(q, J, t)$ to generate a canonical transformation to the new variables $(J, \bar{Q})$, which are again constants of the motion

$$p = \frac{\partial \bar{S}(q, J, t)}{\partial q}$$

$$
\begin{aligned}
\bar{Q} &= \frac{\partial \bar{S}(q, J, t)}{\partial J} \\
&= \frac{\partial \bar{W}(q, J)}{\partial J} - \frac{dE(J)}{dJ}t \equiv \bar{\beta} \qquad \text{; constant}
\end{aligned} \tag{8.22}
$$

---

[33]It is convenient to include the factor $1/2\pi$ in the definition, so that this equation (and subsequent ones based on it) differs from Eq. (36.2) by this factor.

Define the *angle* variable ϕ as[34]

$$\phi = \frac{\partial \bar{W}(q, J)}{\partial J} \qquad (8.23)$$

It follows that

$$\phi = \omega t + \bar{\beta}$$
$$\text{where} \qquad \omega \equiv \frac{dE(J)}{dJ} \qquad (8.24)$$

Thus $\phi(t)$ increases linearly with the time at a rate ω, with $\bar{\beta}$ as the initial value.

Consider the change in the angle variable as the coordinate executes one cycle. As q changes by dq, it follows from Eq. (8.23) that the change in ϕ is given by

$$d\phi = \frac{\partial^2 \bar{W}(q, J)}{\partial q \partial J} dq \qquad (8.25)$$

Hence the net change $\Delta\phi$ around the cycle is

$$\Delta\phi = \oint_{\text{cycle}} d\phi = \oint_{\text{cycle}} \frac{\partial^2 \bar{W}(q, J)}{\partial q \partial J} dq$$
$$= \frac{\partial}{\partial J} \oint_{\text{cycle}} \frac{\partial \bar{W}(q, J)}{\partial q} dq \qquad (8.26)$$

With the identification of p from Eqs. (8.21) and (8.22), one finds

$$\Delta\phi = \frac{\partial}{\partial J} \oint_{\text{cycle}} p \, dq$$
$$= \frac{\partial}{\partial J} 2\pi J = 2\pi \qquad (8.27)$$

On the other hand, we know from Eq. (8.24) that the change in angle variable around the cycle is $\Delta\phi = \omega\tau$ where τ is the period of the motion. Hence

$$\Delta\phi = \omega\tau = 2\pi$$
$$\tau = \frac{2\pi}{\omega} \qquad \text{; period} \qquad (8.28)$$

These results have several important features:

- The angle variable increases *linearly with the time* [Eq. (8.24)]. No matter how complicated the nonlinear dynamics, the angle variable is a *pure rotation*, as defined at the beginning of this section. This conclusion holds not only for the rotational motion of the pendulum but also for the libration, where the angle θ remains bounded;

---

[34]In general, ϕ is not a physical angle in the usual sense.

- Since J is a constant, the trajectory in the (ϕ, J) phase space is, in fact, just a *straight horizontal line*;

- The *angular frequency* of the motion is given in Eq. (8.24),

$$\omega = \frac{dE(J)}{dJ} = \left[\frac{dJ(E)}{dE}\right]^{-1} \tag{8.29}$$

where the latter expression explicitly gives $\omega(E)$, frequently the most useful form;

- The *period* of the motion is given in Eq. (8.28),

$$\tau = \frac{2\pi}{\omega} = 2\pi \frac{dJ(E)}{dE} \tag{8.30}$$

which gives $\tau(E)$ as an explicit function of E;

- If one is simply interested in the period, it remains only to compute $J(E)$ and then $E(J)$;

- It is clear from the derivation that $(J, \bar{Q}) \equiv (J, \bar{\beta})$ are a canonical pair; however, (J, ϕ) also form a canonical pair since $\bar{\beta}$ and ϕ differ only by a linear time shift (see later).

We proceed to analyze our two examples (simple harmonic oscillator and pendulum) in terms of action-angle variables.

Simple harmonic oscillator

Start from the energy and corresponding momentum

$$E = \frac{p^2}{2m} + \frac{1}{2}m\omega_0^2 q^2$$

$$p = \pm\sqrt{2m}\sqrt{E - \frac{1}{2}m\omega_0^2 q^2} \tag{8.31}$$

First compute the action in Eq. (8.19) by integrating around the elliptical trajectory in the clockwise direction. Let q_0 be the maximum extent of the coordinate, so that $E = m\omega_0^2 q_0^2/2$. There will be four identical contributions to J, one from each quadrant, giving

$$J(E) = \frac{4\sqrt{2m}}{2\pi} \int_0^{q_0} dq \sqrt{E - \frac{1}{2}m\omega_0^2 q^2} \quad ; \quad E = \frac{1}{2}m\omega_0^2 q_0^2 \tag{8.32}$$

Let $x = q/q_0$; then

$$J(E) = \frac{2q_0\sqrt{2mE}}{\pi} \int_0^1 dx \sqrt{1 - x^2}$$

$$= \frac{2q_0\sqrt{2mE}}{\pi} \frac{\pi}{4} = \frac{E}{\omega_0} \tag{8.33}$$

64

In this case the inversion is immediate, and we find the obvious result

$$E(J) = \omega_0 J$$
$$\omega = \frac{dE}{dJ} = \omega_0 \tag{8.34}$$

This linear relation between J and E is special to the simple harmonic oscillator.

From Eqs. (8.18) and (8.34), the solution to the H-J equation is (we choose to start with the positive root)

$$W(q, E) = \sqrt{2m} \int^q dq \sqrt{E - \frac{1}{2}m\omega_0^2 q^2} + W_0(E)$$

$$\bar{W}(q, J) = W[q, E(J)] = \sqrt{2m} \int^q dq \sqrt{\omega_0 J - \frac{1}{2}m\omega_0^2 q^2} + \bar{W}_0(J) \tag{8.35}$$

where $\bar{W}_0$ is an integration constant independent of q. The angle variable is thus given by Eq. (8.23) as

$$\phi = \frac{\partial \bar{W}(q, J)}{\partial J}$$

$$= \frac{\omega_0}{2}\sqrt{2m} \int^q dq \frac{1}{\sqrt{\omega_0 J - \frac{1}{2}m\omega_0^2 q^2}} + \phi_0(J) \tag{8.36}$$

Again introduce $x = q/q_0$, with $\omega_0 J = m\omega_0^2 q_0^2/2$, and choose an orientation of the coordinate axes so that $\phi = 0$ when $(p = p_{\max}, q = 0)$. Then

$$\phi = \int_0^x \frac{dx}{\sqrt{1 - x^2}}$$

$$= \sin^{-1} x \qquad ; x = q/q_0 \tag{8.37}$$

where $\sin^{-1} x$ is the inverse sine function $\arcsin x$. Inversion of this relation yields $x = \sin \phi$ or

$$q = \sqrt{\frac{2J}{m\omega_0}} \sin \phi \tag{8.38}$$

The corresponding value of p is obtained from Eq. (8.31)

$$p^2 = 2m\left(\omega_0 J - \frac{1}{2}m\omega_0^2 q^2\right)$$

$$= 2m\omega_0 J(1 - x^2) = 2m\omega_0 J \cos^2 \phi$$

$$p = \sqrt{2m\omega_0 J} \cos \phi \tag{8.39}$$

where the sign is chosen so that now both (p, q) are positive when ϕ is in the first quadrant.

65

In *summary*, the familiar simple harmonic oscillator has the less familiar description in action-angle variables

$$p = \sqrt{2m\omega_0 J}\cos\phi$$

$$q = \sqrt{\frac{2J}{m\omega_0}}\sin\phi$$

$$\phi = \omega_0 t \tag{8.40}$$

Although we have derived these results only in the first quadrant for (q,p), they are now explicity periodic in ϕ in the interval $[0, 2\pi]$ and can thus be extended to the entire interval. The phase-space ellipse in (p,q) is recovered, and ϕ parametrizes the trajectory around the ellipse, starting with $\phi = 0$ at $(p = p_{max}, q = 0)$, moving in the clockwise direction, and increasing linearly with time.[35] In this special case, the angle variable ϕ is just the phase $\omega_0 t + \phi_0$ of the trigonometric variables p and q. Note that the *clockwise* sense of motion means that ϕ is the negative of the conventional polar angle, which increases in the *counterclockwise* direction following the usual right hand rule. In contrast to the closed orbit in (q,p) phase space, the orbit in (ϕ, J) phase space is simply a straight horizontal line that is periodic in ϕ with period 2π; the ϕ coordinate for a simple harmonic oscillator increases linearly with time, and the temporal period τ is given by $\tau = 2\pi/\omega_0$.

The reader will no doubt consider this a rather complicated way to solve the simple harmonic oscillator. The corresponding analysis of the pendulum is considerably more challenging and interesting.

Pendulum

Start from the hamiltonian and corresponding momentum of the planar pendulum

$$H = \frac{p_\theta^2}{2ml^2} + mgl(1 - \cos\theta)$$

$$p_\theta = \pm\sqrt{2ml^2}\sqrt{E - 2mgl\sin^2\chi} \qquad ; \chi \equiv \theta/2 \tag{8.41}$$

where we have introduced the new angular variable $\chi \equiv \theta/2$. The action is then given by

$$J = \frac{1}{2\pi}\oint_{cycle} 2d\chi\sqrt{2ml^2}\sqrt{E - 2mgl\sin^2\chi} \tag{8.42}$$

where $\oint_{cycle}$ contains the appropriate signs. We proceed to calculate J for the two types of motion of the pendulum:

1) *Libration.* Write the energy in terms of the maximum angular displacement of the pendulum $\chi_0 = \theta_0/2$, and introduce the angular frequency of the

---

[35]Note that a rescaling of the coordinates $(p, q) \rightarrow (p/\sqrt{2m\omega_0 J}, q\sqrt{m\omega_0/2J})$ converts the phase space orbit into a circle, with ϕ a polar angle.

simple pendulum

$$E = 2mgl \sin^2 \chi_0 \qquad ; \omega_0 = \sqrt{\frac{g}{l}} \qquad (8.43)$$

Then compute

$$
\begin{aligned}
J &= \frac{2}{2\pi} \oint_{\text{cycle}} d\chi \sqrt{2ml^2} \sqrt{2mgl} \sqrt{\frac{E}{2mgl} - \sin^2 \chi} \\
&= \frac{4ml^2 \omega_0}{2\pi} 4 \int_0^{\chi_0} d\chi \sqrt{\sin^2 \chi_0 - \sin^2 \chi} \qquad (8.44)
\end{aligned}
$$

where we have again noted that there are four equal contributions around the cycle.

To proceed, it is convenient to introduce the new variable

$$
\begin{aligned}
\psi &= \sin \chi \qquad ; d\psi = \cos \chi \, d\chi \\
d\chi &= \frac{d\psi}{\sqrt{1 - \psi^2}} \qquad (8.45)
\end{aligned}
$$

Hence, with $x \equiv \psi/\psi_0$, some algebra yields

$$
\begin{aligned}
J &= \frac{8ml^2 \omega_0}{\pi} \int_0^{\psi_0} d\psi \frac{\sqrt{\psi_0^2 - \psi^2}}{\sqrt{1 - \psi^2}} \\
&= \frac{8ml^2 \omega_0}{\pi} \left[\int_0^1 dx \sqrt{\frac{1 - k^2 x^2}{1 - x^2}} - (1 - k^2) \int_0^1 \frac{dx}{\sqrt{(1 - x^2)(1 - k^2 x^2)}} \right] \\
k^2 &= \psi_0^2 = \sin^2 \chi_0 = \frac{E}{2mgl} < 1 \qquad (8.46)
\end{aligned}
$$

This rather complicated rewriting reduces the integral to a standard form in terms of elliptic integrals of the first and second kind

$$J = \frac{8ml^2 \omega_0}{\pi} [\mathcal{E}(k^2) - (1 - k^2) \mathcal{K}(k^2)] \qquad ; k^2 \equiv \frac{E}{2mgl} \qquad (8.47)$$

Here

$$
\begin{aligned}
\mathcal{K}(k^2) &\equiv \int_0^1 dx \frac{1}{\sqrt{(1 - x^2)(1 - k^2 x^2)}} \\
\mathcal{E}(k^2) &\equiv \int_0^1 dx \sqrt{\frac{1 - k^2 x^2}{1 - x^2}} \qquad (8.48)
\end{aligned}
$$

with $0 \le k^2 < 1$ to ensure convergence of the integral $\mathcal{K}(k^2)$.[36]

---

[36] We here follow the notation of Abramowitz and Stegun [Ab64] and of Mathematica [Wo96] in writing the arguments of the elliptic integrals as k^2. The corresponding Jacobi elliptic function is written as $\text{sn}(y \mid k^2)$. In contrast, most older books use k as the argument. The reader should be careful in comparing results from different sources.

For small k^2 (and thus low energy), the small-amplitude expansion $k^2 \to 0$ gives [Wo05]

$$\mathcal{K}(k^2) \;\to\; \frac{\pi}{2}\left(1 + \frac{k^2}{4} + \cdots\right)$$

$$\mathcal{E}(k^2) \;\to\; \frac{\pi}{2}\left(1 - \frac{k^2}{4} + \cdots\right)$$

$$\mathcal{E}(k^2) - (1-k^2)\,\mathcal{K}(k^2) \;\to\; \frac{\pi k^2}{4} + \cdots \qquad ; \; k^2 \to 0 \qquad (8.49)$$

Hence one recovers the small-amplitude libration result that the pendulum reduces to the simple harmonic oscillator

$$J \;=\; \frac{8ml^2\omega_0}{\pi}\,\frac{\pi k^2}{4} \;=\; \frac{E}{\omega_0} \qquad\qquad ; \; k^2 \to 0 \qquad (8.50)$$

with oscillation frequency $dE/dJ = \omega_0$, as seen in Eq. (8.34).

For general $k^2 < 1$ (or $E < 2mgl$), Eq. (8.47) determines the action $J(E)$. With the following standard formulas [Wo05]

$$\frac{\partial \mathcal{K}}{\partial k^2} \;=\; \frac{1}{2k^2}\left(\frac{\mathcal{E}}{1-k^2} - \mathcal{K}\right)$$

$$\frac{\partial \mathcal{E}}{\partial k^2} \;=\; \frac{\mathcal{E} - \mathcal{K}}{2k^2} \qquad\qquad\qquad (8.51)$$

the anharmonic oscillation frequency of the libration $\omega = dE/dJ = (dJ/dE)^{-1}$ becomes

$$\omega \;=\; \frac{\pi\omega_0}{2\mathcal{K}(k^2)} \qquad\qquad (8.52)$$

As expected, the frequency decreases continuously with increasing $E/(2mgl) = k^2$ and approaches 0 as $k^2 \to 1^-$ from below the separatrix. Figure 8.1 shows both the scaled frequency (ω/ω_0) and the scaled period $(\tau/\tau_0 = \omega_0\tau/2\pi = \omega_0/\omega)$ for the pendulum, including both libration $(E < 2mgl)$ and rotation $(E > 2mgl)$. These same results also follow directly from the period of one cycle as found from the conservation of energy (see problems).

2) *Rotation.* Here $E > 2mgl$ and the angle θ increases continuously from 0 to 2π over one complete cycle. Hence

$$J \;=\; \frac{\sqrt{2ml^2E}}{\pi} \int_0^\pi d\chi \sqrt{1 - k^{-2}\sin^2\chi}$$

$$\;=\; \frac{2\sqrt{2ml^2E}}{\pi} \int_0^{\pi/2} d\chi \sqrt{1 - k^{-2}\sin^2\chi} \qquad ; \; k^{-2} \equiv \frac{2mgl}{E} \quad (8.53)$$

The same manipulations as above reduce this integral to a standard form

$$J \;=\; \frac{2\sqrt{2ml^2E}}{\pi}\,\mathcal{E}(k^{-2}) \qquad\qquad (8.54)$$

where $\mathcal{E}$ is defined in Eq. (8.48). Note that we still define $k^2 = E/(2mgl)$, so that $0 \leq k^{-2} < 1$ for any allowed rotational motion.

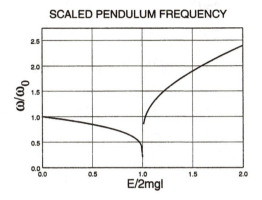

SCALED PENDULUM FREQUENCY

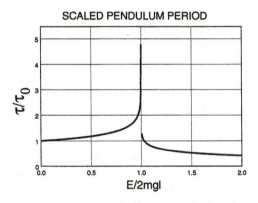

SCALED PENDULUM PERIOD

Fig. 8.1. Anharmonic frequency and period for the pendulum, showing both libration ($E < 2mgl$) and rotation ($E > 2mgl$). The ordinate is the scaled frequency ω/ω_0 or the scaled period $\tau/\tau_0 = \omega_0\tau/2\pi = \omega_0/\omega$, and the abscissa is the scaled energy $k^2 = E/2mgl$. Both branches in the bottom figure diverge near $k^2 = 1$ since $\mathcal{K}(1 - \varepsilon) \to \ln(4/\sqrt{\varepsilon})$ as $\varepsilon \to 0^+$ [Wo05]. Indeed, each branch in either figure is singular near $k^2 = 1$ in the sense that the first derivative diverges.

In particular, the *high-energy* limit is obtained here as $k^{-2} = 2mgl/E \to 0$, with the result

$$J = \frac{2\sqrt{2ml^2E}}{\pi} \frac{\pi}{2} = \sqrt{2ml^2E} \qquad ; k^{-2} \to 0 \qquad (8.55)$$

where we have used Eq. (8.49). This result is simply that for a rigid rotor of moment of inertia ml^2, and inversion gives

$$E = \frac{J^2}{2ml^2}$$

$$\frac{dE}{dJ} = \frac{J}{ml^2} \qquad ; k^{-2} \to 0 \qquad (8.56)$$

If the energy is expressed in terms of the angular frequency $\dot{\theta}$ of rotation, then one finds

$$E = \frac{1}{2}ml^2\dot{\theta}^2$$

$$\omega = \frac{J}{ml^2} = \frac{\sqrt{2ml^2E}}{ml^2}$$

$$= \dot{\theta} \qquad ; k^{-2} \to 0 \qquad (8.57)$$

which is the anticipated high-energy result.

More generally, Eq. (8.54) determines the action $J(E)$ for rotational motion in the allowed range $E > 2mgl$ and hence the rotation frequency for general E. A straightforward calculation gives

$$\omega = \frac{dE}{dJ} = \left(\frac{dJ}{dE}\right)^{-1} = \sqrt{\frac{E}{2ml^2}}\frac{\pi}{\mathcal{K}(k^{-2})}$$

$$\omega = \frac{\pi\omega_0 k}{\mathcal{K}(k^{-2})} \qquad (8.58)$$

As expected from physical considerations, the frequency tends to 0 as $k^2 \to 1^+$ from above the separatrix. In contrast, the frequency increases monotonically with increasing $E/2mgl > 1$ and grows like $\sqrt{E}$ for large E, as seen in Eq. (8.57). Figure 8.1 shows both the scaled frequency ω/ω_0 and the period $\tau/\tau_0 = \omega_0/\omega$ throughout the range of $E/2mgl$.

We note that as $k^2 \to 1$, the coefficient of $1/\mathcal{K}$ in the angular frequency for rotation in Eq. (8.58) is a factor of 2 larger than that for libration in Eq. (8.52).[37] The origin of this factor is easy to understand if one returns to Fig. 1.1. For a libration lying just inside the separatrix, one traverses the entire closed orbit (namely, the pendulum swings in both directions through a complete cycle). In contrast, a rotation lying just above the separatrix traverses just the upper half of an extended (unphysical) libration curve (namely, the pendulum rotates in one direction). To mimic the complete libration, one would have to include the other rotation curve lying just below the lower separatrix, which moves in the opposite direction.

---

[37]For librations with $k^2 = 1 - \varepsilon$, one requires $\mathcal{K}(k^2) = \mathcal{K}(1 - \varepsilon) \approx \ln(4/\sqrt{\varepsilon})$ in the angular frequency as $\varepsilon \to 0$. For rotations with $k^2 = 1 + \varepsilon$, one requires $\mathcal{K}(k^{-2}) \approx \mathcal{K}(1 - \varepsilon)$, which is exactly the same expression.

Now consider the H-J equations for the pendulum

$$\frac{1}{2ml^2}\left(\frac{d\bar{W}}{d\theta}\right)^2 + mgl(1-\cos\theta) = E(J)$$

$$\frac{d\bar{W}}{d\chi} = 2\sqrt{2ml^2}\sqrt{E(J)-2mgl\sin^2\chi}$$

$$\chi = \frac{\theta}{2} \qquad (8.59)$$

where we use the same substitution $\chi = \theta/2$ and again choose to start with the positive root in the second line. Integration gives

$$\bar{W}(\chi, J) = 2\sqrt{2ml^2}\int^\chi d\chi\,\sqrt{E(J)-2mgl\sin^2\chi} + \bar{W}_0(J) \qquad (8.60)$$

The angle variable ϕ follows as

$$\phi = \frac{\partial\bar{W}(\chi, J)}{\partial J} = \frac{d\bar{W}(\chi, J)}{dE}\frac{dE(J)}{dJ}$$

$$= \omega\sqrt{\frac{l}{g}}\int^\chi \frac{d\chi}{\sqrt{E(J)/2mgl - \sin^2\chi}} + \phi_0(J) \qquad (8.61)$$

Here Eq. (8.29) has been used in the second line. We again consider two cases:

1) *Libration.* With $\psi = \sin\chi$ and a coordinate system where $\phi = 0$ when $\psi = 0$, the integral becomes

$$\phi = \omega\sqrt{\frac{l}{g}}\int_0^\psi \frac{d\psi}{\sqrt{\psi_0^2 - \psi^2}\sqrt{1-\psi^2}}$$

$$= \frac{\omega}{\omega_0}\text{sn}^{-1}\left(\psi/\psi_0 \mid k^2\right) \qquad ;\ \phi = \omega t \qquad (8.62)$$

where $\text{sn}^{-1}(y \mid k^2)$ is the inverse Jacobi elliptic function[38]

$$\text{sn}^{-1}(y \mid k^2) = \int_0^y \frac{dx}{\sqrt{1-x^2}\sqrt{1-k^2x^2}} \qquad (8.63)$$

and we have defined

$$\psi \equiv \sin\frac{\theta}{2} \qquad ;\ \psi_0 = \sin\frac{\theta_0}{2}$$

$$k \equiv \sqrt{\frac{E}{2mgl}} \qquad (8.64)$$

In the case of libration where $k^2 < 1$, one has $\psi_0 = k$.

---

[38]This expression has the two evident limiting cases [Wo05], $\text{sn}^{-1}(y|k^2) \to \sin^{-1}y$ as $k^2 \to 0$, and $\text{sn}^{-1}(y|k^2) \to (1/k)\sin^{-1}ky$ as $k^2 \to \infty$, which will prove useful in the subsequent analysis. Note also that $\text{sn}^{-1}(y \mid 1) = \frac{1}{2}\ln[(1+y)/(1-y)] = \tanh^{-1}y$.

Equivalently, we can invert Eq. (8.62) to write

$$\psi(t) \;=\; \sin\frac{\theta(t)}{2} \;=\; k\operatorname{sn}(\omega_0 t \,|\, k^2) \tag{8.65}$$

Equations (8.62) and (8.65) determine how the coordinate amplitude $\psi = \sin\theta/2$ oscillates in time as the system follows the horizontal straight trajectory ($\phi = \omega t$, $J = $ constant) in (ϕ, J) phase space.

For $k^2 \to 0$, one has $\theta \to 0$ and [Wo05] $\operatorname{sn}(y\,|\,0) = \sin y$. Hence, in this limit, Eq. (8.65) gives

$$\theta \;=\; \theta_0 \sin\omega_0 t \qquad\qquad ; k^2 \to 0 \tag{8.66}$$

This is the familiar result for a simple pendulum.

For $k^2 \to 1$, one has $\sin(\theta_0/2) \approx 1$ and the Jacobi sn function now reduces to a hyperbolic tangent [Wo05] $\operatorname{sn}(y\,|\,1) = \tanh y$. Hence, in this limit, Eq. (8.65) gives

$$\sin\frac{\theta}{2} \;=\; \tanh\omega_0 t \qquad\qquad ; k^2 \to 1 \tag{8.67}$$

Here, the pendulum starts vertically downward at rest with energy $E = 2mgl$ and asymptotically approaches the upward position with characteristic time ω_0^{-1}. This behavior agrees with the linearized analysis near the unstable fixed point discussed in connection with Eq. (8.11). Note that the pendulum now takes an essentially infinite time to go from the initial vertical down position to the final vertical up position; correspondingly, the frequency is effectively zero.

The libration amplitude in Eq. (8.65) is plotted as a function of $\omega_0 t$ for two representative values of $k^2 = E/2mgl$ in Fig. 8.2. Note that the amplitude $\psi(t) = \sin[\theta(t)/2]$ never reaches 1 for any $k^2 < 1$.

2) *Rotation*. An analogous calculation in the case of rotation where $k^2 = E/2mgl > 1$ gives

$$\phi \;=\; \frac{\omega}{\omega_0}\int_0^\psi \frac{d\psi}{\sqrt{k^2 - \psi^2}\,\sqrt{1 - \psi^2}}$$

$$\;=\; \frac{\omega}{\omega_0 k}\operatorname{sn}^{-1}\left(\psi\,|\,k^{-2}\right) \qquad\qquad ; \phi \;=\; \omega t$$

or $\qquad \psi(t) \;=\; \sin\frac{\theta(t)}{2} \;=\; \operatorname{sn}(\omega_0 k t\,|\,k^{-2}) \tag{8.68}$

Since $\operatorname{sn}(x\,|\,k^{-2}) \to \sin x$ as $k^2 \to \infty$, one finds

$$\omega t \;=\; \frac{\omega}{\omega_0}\sqrt{\frac{2mgl}{E}}\,\frac{\theta}{2}$$

or $\qquad \theta \;=\; \dot\theta\, t \qquad\qquad ; k^2 \to \infty \tag{8.69}$

where the first of Eqs. (8.57) has been employed in obtaining the last equality. This is again the familiar high-energy result. The rotation amplitude in Eq. (8.68) is plotted as a function of $\omega_0 t$ for two representative values of $k^2 = E/2mgl$ in Fig. 8.3.

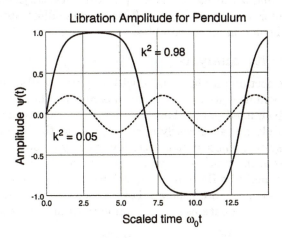

Fig. 8.2. Libration amplitude $\psi(t) = \sin[\theta(t)/2]$ as a function of the scaled time $\omega_0 t$. The two curves correspond to $k^2 = E/2mgl = 0.05$ (dashed) and 0.98 (solid), with frequencies $\omega/\omega_0 \approx 0.987$ and 0.468, respectively, taken from Eq. (8.52). Note the increased period and decreased frequency with increasing $E/2mgl < 1$, as well as the increased anharmonicity of the motion.

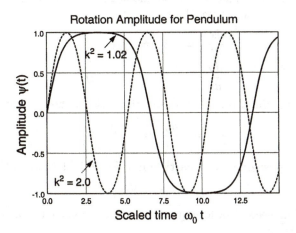

Fig. 8.3. Rotation amplitude $\psi(t) = \sin[\theta(t)/2]$ as a function of the scaled time $\omega_0 t$. The two curves correspond to $k^2 = E/2mgl = 2.0$ (dashed) and 1.02 (solid), with frequencies $\omega/\omega_0 \approx 2.396$ and 0.943, respectively, taken from Eq. (8.58). Note that the absolute value of $\sin(\theta/2)$ now reaches 1 each time the pendulum is vertically upward. In addition, the frequency increases with increasing $E/2mgl > 1$, while the anharmonicity of the motion decreases.

Figures 8.2 and 8.3 show that the motion of the pendulum can be quite complicated, although it is explicitly periodic in time. Nevertheless, the corresponding variables ϕ and J execute a very simple horizontal trajectory in (ϕ, J) phase space, at a uniform speed $d\phi/dt = \omega$. This example illustrates the power and importance of the action-angle variables.

Linearized stability analysis

We conclude this section on phase space with a more detailed discussion of the fixed points of nonlinear dynamical equations and the linearized stability for motion near a given fixed point. This treatment generalizes the analysis of the motion of a pendulum that is nearly upright, given at the beginning of the section. Consider a system whose dynamics is governed by n coupled, first-order differential equations in the time (for a general treatment of such equations, see [Hi74, Ar80]). These might be Hamilton's equations (in which case, n would be even), but the discussion is more general. Form a vector $\boldsymbol{r} \equiv (x_1, x_2, \cdots, x_n)$ from the n variables. The equations can be summarized in the form

$$\frac{d\boldsymbol{r}}{dt} = \dot{\boldsymbol{r}} = \boldsymbol{f}(\boldsymbol{r}, t) \tag{8.70}$$

where $\boldsymbol{f}$ is an n-component function of $\boldsymbol{r}$. It may be interpreted as a *velocity* function for given $(\boldsymbol{r}, t)$. If $\boldsymbol{f}(\boldsymbol{r}, t)$ is independent of t, the equations are said to be *autonomous*. As a function of time, the system traces out some curve $\boldsymbol{r}(t)$ in the n-dimensional configuration ("phase") space that depends on the starting vector $\boldsymbol{r}(0)$.[39]

As an example, consider an autonomous set with $n = 2$. In detail, the equations under discussion have the form

$$\frac{dx}{dt} = \dot{x} = f_x(x, y)$$

$$\frac{dy}{dt} = \dot{y} = f_y(x, y) \tag{8.71}$$

For such a system, the orbit follows from the ratio of these equations

$$\frac{\dot{y}}{\dot{x}} = \frac{dy}{dx} = \frac{f_y(x, y)}{f_x(x, y)} \tag{8.72}$$

For example, suppose one has a particle in a force field $F(x)$ so that

$$m\ddot{x} = F(x) \tag{8.73}$$

Define $y \equiv \dot{x}$. The second-order equation is then converted to a pair of coupled first-order equations

$$\dot{x} = y$$

$$m\dot{y} = F(x) \tag{8.73}$$

---

[39]Here, we generalize the notion of phase space to include n coupled first-order differential equations in the time . For hamiltonian systems, n is even and the structure of Hamilton's equations imposes a very specific form on the function $\boldsymbol{f}$ in Eq. (8.70).

The corresponding orbit equation takes the form

$$mydy = F(x)dx \qquad (8.74)$$

which can be integrated to find $y(x)$ for any given $F(x)$.

Usually one cannot find the orbit $r(t)$ so simply; it is, however, still possible to obtain a few general results. In particular, we can investigate the occurences of *fixed points* where

$$\dot{r} = 0 \qquad ; \text{ fixed points} \qquad (8.75)$$

At these points, the system is *stationary*. Equation (8.70) then implies that these fixed points are characterized by the condition

$$f(r) = 0 \qquad ; \text{ fixed points} \qquad (8.76)$$

Suppose that this set of equations has k real roots $(r_1, r_2, \cdots, r_k)$. For each of these fixed points, we can study the stability of the system in its vicinity. Consider the jth fixed point r_j. If some of the nearby trajectories tend toward r_j, we say that it is an *attractor*. If all nearby trajectories are attracted to it, we say that the jth fixed point is *asymptotically stable*.

To illustrate the method, we return to the example $n = 2$ with two variables x and y. Expand the dynamical equations in the vicinity of this fixed point

$$
\begin{aligned}
x &= x_j + \delta x \\
y &= y_j + \delta y \\
\delta \dot{x} &= \frac{\partial f_x}{\partial x} \delta x + \frac{\partial f_x}{\partial y} \delta y \\
\delta \dot{y} &= \frac{\partial f_y}{\partial x} \delta x + \frac{\partial f_y}{\partial y} \delta y
\end{aligned}
\qquad (8.77)
$$

Here the derivatives are evaluated at r_j, and only the first terms in the Taylor series expansion about the fixed point have been retained. We assume that all quantities are real. These equations can be recast in matrix form as

$$\delta \dot{r} = \underline{A} \, \delta r \qquad (8.78)$$

where

$$
\delta r = \begin{pmatrix} \delta x \\ \delta y \end{pmatrix} \qquad ; \delta \dot{r} = \begin{pmatrix} \delta \dot{x} \\ \delta \dot{y} \end{pmatrix}
$$

$$
\underline{A} = \begin{pmatrix} \partial f_x/\partial x & \partial f_x/\partial y \\ \partial f_y/\partial x & \partial f_y/\partial y \end{pmatrix}
\qquad (8.79)
$$

Take linear combinations $(\delta X, \delta Y)$ of $(\delta x, \delta y)$ of the form

$$\delta R = \underline{M}^{-1} \delta r \qquad (8.80)$$

where

$$\delta R = \begin{pmatrix} \delta X \\ \delta Y \end{pmatrix} \tag{8.81}$$

and $\underline{M}^{-1}$ is a real, constant 2×2 matrix. Upon substitution of Eq. (8.78), the time derivative of Eq. (8.80) yields

$$\delta \dot{R} = \underline{M}^{-1} \underline{A} \underline{M} \delta R \tag{8.82}$$

Now choose the matrix $\underline{M}$ to diagonalize $\underline{A}$, which can be done using the technique in Sec. 22. First find the eigenvalues of $\underline{A}$

$$\det|\underline{A} - \lambda \underline{I}| = 0 \tag{8.83}$$

Here, this real quadratic equation for λ has either real, or complex conjugate roots. The matrix $\underline{M}$ that diagonalizes $\underline{A}$ is just the modal matrix formed from the eigenvectors as columns. Thus $\underline{A}$ acquires diagonal form $\underline{A}_D$

$$\underline{M}^{-1} \underline{A} \underline{M} = \underline{A}_D = \begin{pmatrix} \lambda_1 & 0 \\ 0 & \lambda_2 \end{pmatrix} \tag{8.84}$$

where (λ_1, λ_2) are the eigenvalues of $\underline{A}$.

The solution to Eqs. (8.82) then takes the form[40]

$$\delta R(t) = \begin{bmatrix} \delta X(t) \\ \delta Y(t) \end{bmatrix} = \begin{bmatrix} \delta X(0)e^{\lambda_1 t} \\ \delta Y(0)e^{\lambda_2 t} \end{bmatrix} \tag{8.85}$$

The eigenvalues (λ_1, λ_2) thus describe the linear stability about the fixed point. If the eigenvalues are real,[41] then the solutions will grow or decay according to the signs of (λ_1, λ_2). With complex conjugate eigenvalues of the form $\alpha \pm i\omega$, the solutions will oscillate at frequency ω, and the magnitude of the solution will grow or decay depending on the sign of α. In terms of the original variables δr (the small displacement from the fixed point), the general solution is a linear combination of the various exponentials (here, $e^{\lambda_1 t}$ and $e^{\lambda_2 t}$).

As an example, consider the following (dimensionless) second-order nonlinear differential equation

$$\ddot{x} = \dot{x}^2 - x^2 + x \tag{8.86}$$

As before, this second-order equation can be converted to two coupled first-order equations

$$\begin{aligned} \dot{x} &= y = f_x(x, y) \\ \dot{y} &= y^2 - x^2 + x = f_y(x, y) \end{aligned} \tag{8.87}$$

---

[40] Alternatively, one can simply write

$$r(t) = e^{\underline{A}t} r(0)$$

$$e^{\underline{A}t} = 1 + \underline{A}t + \frac{1}{2}\underline{A}^2 t^2 + \cdots$$

[41] It is proved in Sec. 22 that if $\underline{A}$ is symmetric, the eigenvalues must be real.

The fixed points are located by equating the right hand side of these equations to zero

$$
\begin{aligned}
y &= 0 \\
y^2 - x^2 + x &= 0
\end{aligned}
\tag{8.88}
$$

There are two solutions $(x, y) = (0, 0)$ and $(x, y) = (1, 0)$. The matrix $\underline{A}$ in Eq. (8.79) is readily computed at the general point (x, y) to be

$$
\underline{A}(x, y) = \begin{pmatrix} 0 & 1 \\ 1 - 2x & 2y \end{pmatrix}
\tag{8.89}
$$

At the origin, the eigenvalues are $\lambda = \pm 1$. Hence one of the solutions grows like e^t, indicating that the origin is an unstable fixed point. At the point $(1, 0)$, the eigenvalues are $\lambda = \pm i$. Hence the solutions oscillate like $e^{\pm it}$, indicating that while this fixed point is stable, it is not attracting.[42]

With this background, we proceed to an analysis of the three coupled, first-order, nonlinear Lorenz equations derived in Part II.

9 Lorenz model

In this section we examine the solutions to the Lorenz Eqs. (5.41) and (6.25)

$$
\begin{aligned}
\dot{x} &= -Px + Py \\
\dot{y} &= -y + rx - xz \\
\dot{z} &= -bz + xy
\end{aligned}
\tag{9.1}
$$

where we focus on these nonlinear first-order differential equations as illustrations of more general nonlinear phenomena. Nevertheless, it is helpful to recall that the variables (x, y, z) have a direct physical interpretation, as discussed in the model configuration of Sec. 6 [see Eqs. (6.24) and Fig. 6.1], where $x(t)$ is the flow rate and $y(t)$ and $z(t)$ are temperature variations at two different angular positions. Alternatively, for the Rayleigh-Bénard problem discussed in Sec. 5, $x(t)$ is the amplitude of the lowest spatial harmonic of the convective velocity whereas $y(t)$ and $z(t)$ are amplitudes of two different spatial harmonics of the nonlinear convective temperature (the deviation from the conductive solution, which has a uniform gradient). In addition, r is the dimensionless scaled Rayleigh number, which serves as the control parameter that characterizes the external applied "stress" on the system.

For numerical analysis, we must also choose specific values for the dimensionless parameters (P, b). The quantity b is a function of the geometry, and here we use Eq. (5.43) with $b = 8/3$. The other dimensionless parameter is the Prandtl number $P = \nu/\kappa$, the ratio of the kinematic viscosity ν to the thermal diffusivity κ. Since the critical Rayleigh number in Sec. 5 is independent of P,

---

[42] We leave it as a problem to find the behavior at a general point near the fixed points in Fig. 1.1.

our results will not be particularly sensitive to this quantity, and we somewhat arbitrarily choose $P = 10$. Thus we take[43]

$$P \;=\; 10 \qquad\qquad ; \, b = 8/3 \qquad\qquad (9.2)$$

which are the parameters studied by Lorenz [Lo63].

With these specific numbers, the numerical results nicely illustrate the points that we want to make. Near the onset of convection at the convective instability, they reproduce the behavior studied in Sec. 4, but they also make good physical sense for much larger values of the Rayleigh number, where the analytical theory becomes impossible. In addition, we can readily compare with the results of other studies (see, for example, [Yo85]). Existing mathematics packages on PCs are sufficiently powerful that readers can carry out investigations of other parameter sets on their own. Thus we shall study the following equations:

$$\dot{x} \;=\; -10x + 10y$$
$$\dot{y} \;=\; -y + rx - xz$$
$$\dot{z} \;=\; -\frac{8}{3}z + xy \qquad\qquad (9.3)$$

The only parameter remaining in these Lorenz equations is the effective Rayleigh number r given in Eqs. (5.42) and (6.23).

These coupled, nonlinear, differential equations can be solved numerically by simply picking a set of starting values (x_0, y_0, z_0) and stepping forward in the (rescaled) time.[44] We shall subsequently do so. First, however, we use the analysis of the preceeding section to obtain some analytical results. We find the fixed points and carry out a linearized stability analysis about each fixed point.

Stationary solutions and fixed points

The stationary fixed points are defined by

$$\dot{x} \;=\; \dot{y} \;=\; \dot{z} \;=\; 0 \qquad\qquad (9.4)$$

Setting the right hand side of Eqs. (9.3) equal to zero, one obtains the conditions

$$x \;=\; y$$
$$y \;=\; rx - xz$$
$$z \;=\; \frac{3}{8}xy \qquad\qquad (9.5)$$

There are two sets of solutions:

1. The static conductive state at the origin with

$$x \;=\; y \;=\; z \;=\; 0 \qquad\qquad (9.6)$$

---

[43] As noted in a problem, however, the chaotic behavior that we exhibit occurs only for $P > b+1$ [Lo63]. Fluids with small Prandtl numbers do, in fact, behave quite differently (see, for example, [Mc75, Cr93]). Note that the present choice of b differs from that in Eq. (6.25), where we had $b = 1$.

[44] See Eqs. (5.38) and (6.23) for the definition of the dimensionless time.

2. The steady convective state defined by

$$z = r - 1$$
$$x = y = \pm\sqrt{8(r-1)/3} \tag{9.7}$$

We proceed to make a linearized stability analysis for each solution.

Linearized stability analysis

Linearization of Eqs. (9.3) about the fixed point at the origin in Eqs. (9.6) leads to

$$\delta\dot{x} = -10\,\delta x + 10\,\delta y$$
$$\delta\dot{y} = -\delta y + r\,\delta x$$
$$\delta\dot{z} = -\frac{8}{3}\,\delta z \tag{9.8}$$

We seek a solution where all quantities have a time dependence $\sim e^{\lambda t}$. Substitution then leads to the linear, homogeneous, eigenvalue equations

$$(\lambda + 10)\,\delta x - 10\,\delta y = 0$$
$$-r\,\delta x + (\lambda + 1)\,\delta y = 0$$
$$(\lambda + 8/3)\,\delta z = 0 \tag{9.9}$$

These equations will have a nontrivial solution for the variations only if the determinant of the coefficients vanishes, which leads to

$$(\lambda + 8/3)\,[\lambda^2 + 11\lambda + 10(1-r)] = 0$$
$$\lambda_1 = -8/3$$
$$\lambda_\pm = \frac{1}{2}\left(-11 \pm \sqrt{81 + 40r}\right) \tag{9.10}$$

All three roots are real and negative, provided that

$$r < 1 \qquad\qquad ; \text{ stability at origin} \tag{9.11}$$

This condition implies that the uniform heat-conduction state is *stable* for $r < 1$. In addition, the critical Rayleigh number for the onset of instability of the conductive solution (the fixed point at the origin) is $r_c = 1$, because the eigenvalue λ_+ becomes positive for $r > 1$. This unstable mode is the one that grows exponentially with time in the linearized region, although the (positive) growth rate is small for small positive values of $r - 1$.

If $r > 1$, the instability of the conductive state leads to a new stationary convective state described by the second fixed point found above in Eqs. (9.7)

$$z_{\text{conv}} = r - 1$$
$$x_{\text{conv}} = y_{\text{conv}} = \pm\sqrt{8(r-1)/3} \tag{9.12}$$

79

r	λ_1	λ_2	λ_3
1.000	0.000	-2.667	-11.000
1.100	-0.198	-2.443	-11.026
1.345	-1.233	-1.346	-11.088
1.347	-1.289 + 0.084i	-1.289 - 0.084i	-11.088
2.000	-1.212 + 1.810i	-1.212 - 1.810i	-11.242
10.00	-0.595 + 6.170i	-0.595 - 6.170i	-12.476
24.00	-0.023 + 9.490i	-0.023 - 9.490i	-13.622
25.00	+0.008 + 9.672i	+0.008 - 9.672i	-13.683
100.0	+1.158 + 18.14i	+1.158 - 18.14i	-15.983

Table 9.1: Some representative eigenvalues for the linearized stability analysis around the fixed point representing the stationary convective state as a function of the parameter r. They are obtained as solutions to Eq. (9.14).

Note the "broken symmetry" associated with the $\pm$ sign. A given convective roll can rotate in either of two directions. Additional physical information is needed to choose which solution is appropriate. In this convective regime, the fluid velocity x_{conv} is explicitly nonzero. Equations (9.3) are readily linearized about this new fixed point. If one again seeks a solution where all amplitudes have the time dependence $\sim e^{\lambda t}$, one is led to

$$(\lambda + 10)\,\delta x - 10\,\delta y = 0$$
$$-\delta x + (\lambda + 1)\,\delta y \pm \sqrt{8(r-1)/3}\,\delta z = 0$$
$$\mp\sqrt{8(r-1)/3}\,\delta x \mp \sqrt{8(r-1)/3}\,\delta y + (\lambda + 8/3)\,\delta z = 0 \qquad (9.13)$$

These linear, homogeneous, algebraic equations will again have a nontrivial solution for the variations only if the determinant of the coefficients vanishes. A little algebra leads to the eigenvalue equation

$$f(\lambda, r) \equiv \lambda^3 + \frac{41}{3}\lambda^2 + \left(\frac{80}{3} + \frac{8}{3}r\right)\lambda + \frac{160}{3}(r-1) = 0 \qquad (9.14)$$

Inspection of $f(\lambda, r)$ shows that it approaches $\pm\infty$ for $\lambda \to \pm\infty$, so that there is necessarily one real root for any r. If $r = 1$, one root is 0 and the other two are negative (-11 and $-8/3$). For small positive $r - 1$, $f(\lambda, r)$ has three negative roots. As r increases, however, the whole cubic curve rises, and the local minimum with the two rightmost negative roots eventually crosses the real axis. For this special value of r, these two roots become degenerate, and they then move into the complex plane as a complex conjugate pair for larger r.

This cubic equation $f(\lambda, r) = 0$ can be solved numerically for any r, and the three solutions for some characteristic values of r are given in Table 9.1.[45] One observes

---

[45]Note if one finds a complex root λ_1, then $\lambda_2 = \lambda_1^*$ also provides a solution.

- For $1 < r \approx 1.346$, the cubic equation has three negative real roots whose positions depend on r, implying stability of this convective state. Small variations around the convective state will decay exponentially back to it.

- As r increases through the value ≈ 1.346, the two real roots (λ_1 and λ_2) nearest the origin become degenerate and then move into the complex plane as a complex conjugate pair $\lambda_r \pm i\lambda_i$.

- For larger values of r between ≈ 1.346 and ≈ 24.7, the eigenvalue λ_3 remains real and negative; the real part λ_r of the other two eigenvalues also remains negative (although these latter two roots have imaginary parts $\pm\lambda_i$). In this range of r, the convective fixed point is stable, and small variations will oscillate back to it with decaying amplitude (Fig. 9.1);

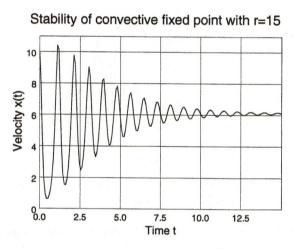

Fig. 9.1. Illustration of stability of convective fixed point with $r = 15$. From numerical solution to Lorenz Eqs. (9.3) with $(x_0, y_0, z_0) = (10, 10, 22)$.

- As shown in the problems, the cubic equation can be solved exactly to determine the point beyond which small perturbations around the convective solution start to grow exponentially (namely, where $\lambda_r = 0$). Straightforward analysis [Lo63] yields the critical scaled Rayleigh number $r^* = P(P+b+3)/(P-b-1) = 470/19 \approx 24.737$ for the present parameters $b = 8/3$ and $P = 10$. At this critical Rayleigh number, the corresponding eigenvalues are $\pm i\lambda_i = \pm i\sqrt{2bP(P+1)/(P-b-1)} \approx \pm 9.625\,i$ and $\lambda_3 = -(P+b+1) = -41/3 \approx -13.667$.

- For r greater than $r^* \approx 24.7$, two of the eigenvalues have a positive real part, implying that the convective state is *unstable*, namely that the instability will grow (Fig. 9.2). The linearized analysis predicts how the small variations will grow initially (but note in Fig. 9.2 that the velocity eventually even reverses direction). This instability indicates a transition from the steady convective rolls found in Sec. 4 to ones that are time dependent. Such behavior is not surprising, because there are no stable fixed points in this regime of r. Analytical studies of this transition are intricate because they involve small perturbations of a state with spatial periodicity, similar to the Bloch states for electrons in a one-dimensional periodic crystal lattice.

- For values of $r \gg 24.7$, our linearized stability analysis about the fixed points provides little useful information.

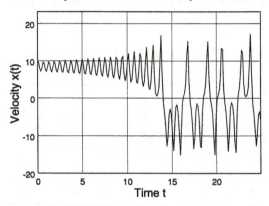

Fig. 9.2. Illustration of instability of convective fixed point with $r = 28$. From numerical solution to Lorenz Eqs. (9.3) with $(x_0, y_0, z_0) = (10, 10, 27)$.

We observe in Fig. 9.2 that the instability soon turns into apparently random motion, with the velocity oscillating around a mean flow and then repeatedly reversing its overall direction. The motion is not random, of course, since the trajectory follows reproducibly from the given set of initial conditions. As the time increases, the exact pattern becomes more and more sensitive to these starting conditions. One might expect to see this behavior become even more pronounced for very large r; however, there are some surprises. For certain intervals of large r, one finds fascinating, rather complex, periodic motion!

Periodic oscillatory solutions

Upon exploration of the numerical integration of the Lorenz equations for intervals of large r, regularities can appear out of the apparently random motion. Figure 9.3 shows the velocity $x(t)$ as a function of t for $r = 148.5$. In this case, one observes periodic motion where the velocity undergoes a fascinating, regular, reversal and oscillation. Since the solution $\mathbf{x}(t) = [x(t), y(t), z(t)]$ is a vector in the (here) three-dimensional phase space, one can actually plot the periodic orbit in that space. It can also be investigated through the projection of the orbit onto a two-dimensional plane, and Fig. 9.4 shows that projection on the (z, x) plane. The rather involved orbit shows a lovely regularity.[46]

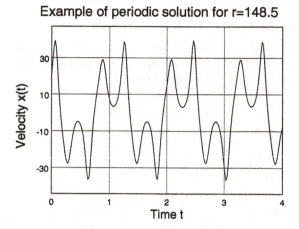

Fig. 9.3. Illustration of periodic oscillatory solution with $r = 148.5$. From numerical solution to Lorenz Eqs. (9.3) with $(x_0, y_0, z_0) = (20, 54, 104)$.

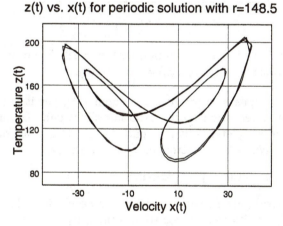

Fig. 9.4. Plot of $z(t)$ vs. $x(t)$ for periodic oscillatory solution with $r = 148.5$. From numerical solution to Lorenz Eqs. (9.3) with $(x_0, y_0, z_0) = (20, 54, 104)$; integrated to $t = 4$.

If the value of r is decreased slightly to $r = 147.5$, close inspection shows that the orbit is actually a union of two congruent orbits slightly separated in (x, z) space, and it takes just twice as long for the system to complete one full orbit — the *period is doubled*. The change in orbit is not dramatic, but readers can

---

[46]This orbit is stable against small deviations from the initial conditions. To get a nice plot, we have started it off near one of its stable values.

investigate this period doubling for themselves.[47] Rather than provide more graphs here, we prefer to wait until the next section where a simple model illustrates much of this behavior in a more transparent fashion.

Chaotic solutions

Let us lower the value of r back down to $r = 28$ (Fig. 9.2). A recalculation of the numerical solution for the velocity $x(t)$ with a slightly different set of initial conditions yields the result shown in Fig. 9.5.[48]

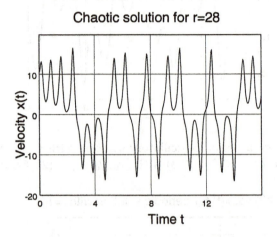

Fig. 9.5. Illustration of chaotic solution with $r = 28$. From numerical solution to Lorenz Eqs. (9.3) with $(x_0, y_0, z_0) = (10, 10, 18)$.

This solution appears to produce a random $x(t)$, yet it is reproduced with a finer and finer integration mesh. Another slight shift in initial conditions, however, yields a completely different pattern. For this value of r, the solution is chaotic, where *chaos* is characterized by a trajectory:

- that is never stationary or periodic;

- whose general pattern with reversals and oscillations does not depend on the initial conditions or integration technique;

- whose detailed structure depends crucially on both;

- that appears to be completely random, yet follows by integration from a given set of initial conditions;

- that is extremely sensitive to those initial conditions.

---

[47] As we shall show in the next section, the period can be most readily extracted from a temporal Fourier transform of the motion.

[48] In the graphs, we are currently limited by the number of plotting points available; the underlying numerical calculations are carried out to high accuracy.

This sensitivity to initial conditions is quantified by the "Liapunov exponent" that characterizes how two initially nearby trajectories evolve in time [Jo98b, Ot02c] (they converge or diverge exponentially according to the sign of the Liapunov exponent).

Although the velocity $x(t)$ appears to have a completely erratic behavior, a plot of the corresponding phase orbit of the temperature $z(t)$ against the velocity $x(t)$ reveals a certain regularity, as shown in Fig. 9.6. The phase orbit is collapsing to a small region in phase space where the temperature and velocity are locked together. We proceed to prove two theorems on the solutions to the Lorenz equations that help one to understand this behavior.

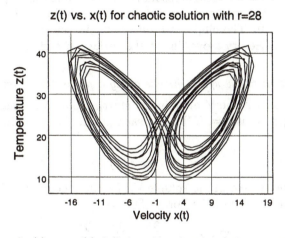

Fig. 9.6. Plot of $z(t)$ vs. $x(t)$ for chaotic solution with $r = 28$. From numerical solution to Lorenz Eqs. (9.3) with $(x_0, y_0, z_0) = (10, 10, 18)$; integrated to $t = 16$.

Two theorems on phase-space convergence of solutions

We prove two theorems on solutions to the Lorenz equations [Sp82]:

1) *The phase point* $\mathbf{x}(t)$ *eventually enters a bounded ellipsoid* $E(x, y, z)$.

To prove this assertion, introduce the *Liapunov function* $L(x, y, z)$

$$L(x, y, z) = rx^2 + Py^2 + P(z - 2r)^2 \tag{9.15}$$

For any given value of L, this quadratic form defines an ellipsoid in (x, y, z) phase space centered at $x = y = 0$, $z = 2r$.

Compute the total time derivative of L with the aid of the Lorenz Eqs. (9.1)

$$
\begin{aligned}
\frac{dL}{dt} &= 2rx\dot{x} + 2Py\dot{y} + 2P(z - 2r)\dot{z} \\
&= 2rx[P(y - x)] + 2Py(rx - y - xz) + 2P(z - 2r)(xy - bz) \\
&= -2P[rx^2 + y^2 + b(z - r)^2 - br^2] \tag{9.16}
\end{aligned}
$$

85

Define another ellipsoid $D(x, y, z)$ by[49]

$$rx^2 + y^2 + b(z - r)^2 = br^2 \tag{9.17}$$

where the right hand side is a constant independent of time. For any point (x, y, z) outside of D, the right hand side of Eq. (9.16) is negative and thus

$$\frac{dL}{dt} < 0 \qquad ; (x, y, z) \text{ outside of } D \tag{9.18}$$

In this region, the numerical value of the quadratic form that defines the ellipsoid $L(x, y, z)$ decreases with time.

Let $E(x, y, z)$ be the smallest ellipsoid L that contains the ellipsoid D. Then a phase trajectory that starts at any point $\mathbf{x}(t)$ which lies on L and outside of E satisfies $dL/dt < 0$. The ellipsoid L on which the phase orbit $\mathbf{x}(t)$ lies then shrinks until the trajectory moves inside the surface E and stays there. This is the first theorem.

2) *The volume in phase space shrinks along a phase trajectory.*

Consider a small volume V of N initial starting points in phase space. The mean density of points is $n = N/V$. For each point, the instantaneous *velocity* is

$$\mathbf{v} = (\dot{x}, \dot{y}, \dot{z}) \tag{9.19}$$

which gives a velocity field in the volume V. The divergence of this velocity can be calculated with the aid of the Lorenz Eqs. (9.1)

$$\nabla \cdot \mathbf{v} = \frac{\partial \dot{x}}{\partial x} + \frac{\partial \dot{y}}{\partial y} + \frac{\partial \dot{z}}{\partial z}$$
$$= -(P + b + 1) \tag{9.20}$$

Now the total number of points is conserved, and hence there is a continuity equation for the number density[50]

$$\frac{\partial n}{\partial t} + \nabla \cdot (n\mathbf{v}) = 0 \tag{9.21}$$

Thus the total, or substantive, derivative of n is given by

$$\frac{dn}{dt} = \frac{\partial n}{\partial t} + \mathbf{v} \cdot \nabla n = -n(\nabla \cdot \mathbf{v}) \tag{9.22}$$

It follows from $V = N/n$ with fixed N that

$$\frac{1}{V}\frac{dV}{dt} = -\frac{1}{n}\frac{dn}{dt}$$
$$= \nabla \cdot \mathbf{v} = -(P + b + 1) \tag{9.23}$$

---

[49]Note that the ellipsoids $D(x, y, z)$ and $L(x, y, z)$ have different shapes and different centers. Recall also that here r is the scaled Rayleigh number.

[50]Just as with charge and current conservation in electricity and magnetism or mass conservation in hydrodynamics.

Integration of this relation yields

$$V(t) = V(0)e^{-(P+b+1)t} \qquad (9.24)$$

This is the stated result. The phase space volume shrinks with time along a phase trajectory. Note that this theorem says nothing about the *shape* of the volume element. Indeed, the chaotic (exponential) divergence of nearby trajectories means that the shape of the volume element must change dramatically for large times [Ot02c].

We have actually observed this behavior in all our examples. The phase volume shrinks in the case of:

- asymptotically stable conductive or convective fixed points where all nearby orbits flow into the fixed point;

- stable periodic orbits that eventually traverse a given curve in phase space;

- chaotic orbits that collapse to lie on a surface inside the bounded ellipsoid in phase space.

In the last case, the surface is often said to form a *strange attractor*.[51]

10 Model finite-difference equation: logistic map

We have seen that the nonlinear Lorenz equations exhibit many fascinating phenomena such as stable and unstable fixed points, periodic solutions, period doubling, and chaos. Most of these properties are generic to nonlinear problems. To illustrate and clarify this behavior, we turn to a very simple model finite-difference equation known as the *logistic map*.

Consider the following first-order, nonlinear differential equation in a single variable

$$\frac{dN}{dt} = \lambda N - \mu N^2 \qquad (10.1)$$

---

[51] A proof exactly analogous to that above yields Liouville's theorem in the case of hamiltonian dynamics. For simplicity, consider a single particle in three dimensions. In the six-dimensional phase space of (x, y, z, p_x, p_y, p_z), the velocity of the point is given by

$$\mathbf{v} = (\dot{x}, \dot{y}, \dot{z}, \dot{p}_x, \dot{p}_y, \dot{p}_z)$$

The divergence of this velocity follows with the aid of Hamilton's equations as

$$\boldsymbol{\nabla} \cdot \mathbf{v} = \sum_i \left(\frac{\partial \dot{x}_i}{\partial x_i} + \frac{\partial \dot{p}_i}{\partial p_i} \right)$$

$$= \sum_i \left\{ \frac{\partial}{\partial x_i} \left(\frac{\partial H}{\partial p_i} \right) - \frac{\partial}{\partial p_i} \left(\frac{\partial H}{\partial x_i} \right) \right\} = 0$$

The same proof then leads to

$$\frac{dV}{dt} = 0$$

Hamiltonian dynamics thus conserves the phase-space volume along a phase trajectory — this is Liouville's theorem. We return to this topic in a later section.

where (λ, μ) are constants, both assumed here to be positive. If $\mu = 0$, this is an elementary growth equation with solution $N = N(0)e^{\lambda t}$. If $\mu \neq 0$, the second term serves to limit the growth as N approaches the value λ/μ from below. With the definition $N = (\lambda/\mu)\, x$, this equation can be rewritten as

$$\frac{dx}{dt} = \lambda x(1 - x) \qquad ; N = \left(\frac{\lambda}{\mu}\right) x \qquad (10.2)$$

Evidently, there are two fixed points, at $x = 0$ and $x = 1$. Linear stability analysis shows that $x = 0$ is an unstable fixed point with solutions that grow like $e^{\lambda t}$. In contrast, $x = 1$ is a stable fixed point with solutions that decay like $\delta x(t) = x(t) - 1 \approx \delta x(0)e^{-\lambda t}$ for small $\delta x(0)$.

In fact, this nonlinear ordinary differential Eq. (10.2) is sufficiently simple that it can be solved exactly. Rearrange the equation as

$$\left(\frac{1}{x} + \frac{1}{1 - x}\right) dx = \lambda\, dt \qquad (10.3)$$

Given an initial value $x(0)$ between 0 and 1, the nonlinear differential Eq. (10.2) is then readily integrated to give

$$x(t) = \frac{x(0)e^{\lambda t}}{1 + x(0)(e^{\lambda t} - 1)} \qquad (10.4)$$

This exact solution has several interesting features.

1. For small $x(0) \ll 1$, the solution initially grows exponentially

$$x(t) \approx x(0)e^{\lambda t} + O[x(0)^2] \qquad (10.5)$$

   This growth illustrates the unstable behavior near the fixed point $x = 0$;

2. As λt increases, the exponential growth saturates because of the denominator in Eq. (10.4). This behavior reflects the factor $1 - x$ in Eq. (10.2);

3. For all initial values $0 < x(0) < 1$, the solution approaches 1 from below, with

$$x(t) \sim 1 - \left[x(0)^{-1} - 1\right] e^{-\lambda t} \qquad (10.6)$$

   for $\lambda t \gg 1$. If $x(0) = 1 - \varepsilon$ is close to 1 with $\varepsilon \ll 1$, this solution $1 - \varepsilon e^{-\lambda t}$ illustrates the behavior near the stable fixed point $x = 1$.

Logistic map

Let us convert Eq. (10.1) to a finite-difference equation, which is the way it is actually integrated numerically

$$
\begin{aligned}
N_{j+1} &= N_j + \left(\lambda N_j - \mu N_j^2\right) \Delta t \qquad ; j = 0, 1, \cdots, n - 1 \\
N_0 &= N(0)
\end{aligned}
\qquad (10.7)
$$

Now define

$$(\mu \, \Delta t) N_j \;\; \equiv \;\; (1 + \lambda \, \Delta t) \, x_j \;\; \equiv \;\; \rho \, x_j \tag{10.8}$$

Then Eq. (10.7) takes the form

$$
\begin{aligned}
x_{j+1} &= \rho \, x_j (1 - x_j) & &; \; j = 0, 1, \cdots, n-1 \\
x_0 &= x(0) & & \tag{10.9}
\end{aligned}
$$

This simple two-term recursion relation is known as the *logistic map*. Its iterated solution, considered as a function of the parameter ρ, has a much richer and more complex behavior than the solution to the corresponding differential equation [they become equivalent only in the limit $(\Delta t = t/n \to 0)$].

Power spectrum

To examine the properties of the iterated solution, it is convenient to focus on the function $x(t)$ defined at the discrete set of points $(x_0, x_1, \cdots, x_n)$

$$x(t) \;\; = \;\; \sum_{m=0}^{n} x_m \delta \, (t - m) \tag{10.10}$$

The Fourier transform of $x(t)$ is given by

$$\int_{-\infty}^{\infty} dt \, x(t) \, e^{i\omega t} \;\; = \;\; \sum_{m=0}^{n} x_m \, \exp(im\omega) \tag{10.11}$$

and it will be convenient to use the conventional frequency ν instead of the angular frequency $\omega = 2\pi\nu$. The quantity $P(\nu)$ is defined by the absolute value squared of the Fourier transform

$$P(\nu) \;\; = \;\; \frac{1}{T} \left| \int_{-\infty}^{\infty} x(t) e^{i\omega t} dt \right|^2 \qquad ; \; \omega = 2\pi\nu \tag{10.12}$$

where the total time interval in Eq. (10.10) is $T = n$. This definition gives

$$P(\nu) \;\; = \;\; \frac{1}{T} \left| \sum_{m=0}^{n} x_m \exp\left(2\pi i \nu \, m\right) \right|^2 \tag{10.13}$$

The function $P(\nu)$ is said to provide the *power spectrum* of the solution $x(t)$. Clearly, it probes the frequencies ν and corresponding periods $\tau = 1/\nu$ contained in the function $x(t)$.

In fact, we first calculate the complex quantity

$$
\begin{aligned}
c_k &= \frac{1}{\sqrt{n+1}} \sum_{m=0}^{n} x_m \exp\left[2\pi i \left(\frac{k}{n+1} \right) m \right] \\
k &= 1, 2, \cdots, \frac{1}{2}(n+1) \tag{10.14}
\end{aligned}
$$

89

which is proportional to the Fourier transform of $x(t)$ evaluated at the discrete frequencies $\nu = k/(n+1)$. What we shall plot are the quantities $|c_k|^2$. To use the common and efficient numerical method known as the fast Fourier transform (FFT), the integer $(n+1)$ must be an integer power of 2, of the form[52]

$$n + 1 = 2^q \qquad\qquad ; q \text{ integer} \qquad\qquad (10.15)$$

This power spectrum probes the frequencies $\nu = k/(n+1)$ and periods $\tau = 1/\nu = (n+1)/k$ of the iterated sequence. Thus if the solution in Eq. (10.10) repeats itself after every second term, it has a strong $|c_k|^2$ at a period 2 and frequency 1/2; if it repeats only after every fourth term, it has a strong $|c_k|^2$ at a period 4 and frequency 1/4, *etc.*

Numerical analysis

It is the simplest of tasks to iterate Eqs. (10.9) on a PC. We proceed to show some representative numerical results. The iterated sequences for values of the parameter $\rho = 2$ and $\rho = 3.3$ are shown in Figs. 10.1(a) and 10.1(b).

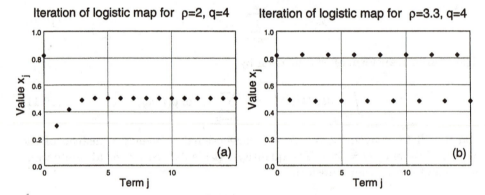

Figs. 10.1(a, b). Plot of x_j vs. j for iteration of the logistic map for: (a) $\rho = 2.0$; and (b) $\rho = 3.3$. Here $(x_0, q) = (0.82, 4)$, where q determines the number of points in the sequence.[53]

In the first case, the iterated solution quickly approaches a fixed value $x_j = 0.5$. In the second case, the solution performs periodic oscillation between $x_j = 0.82$ and $x_j = 0.48$, with a period of $\tau = 2$.

The result of increasing the value of ρ slightly to $\rho = 3.5$ is shown in Fig. 10.2(a). There is now a four-cycle oscillation between the values $(0.88, 0.38, 0.83, 0, 50)$ with period of $\tau = 4$. The period has doubled. If the value of ρ is increased again to $\rho = 4$, the result is shown in Fig. 10.2(b). The iterated values of x_j appear to be completely random — the situation is chaotic.

---

[52]In Mathcad11, the expression in Eq. (10.14) is calculated as a fast Fourier transform $fft(x)$ of the data set $\{x\} = (x_0, x_1, \cdots, x_n)$.
[53]The results shown in these two figures, as well as in Fig. 10.2(a), are stable against changes in the starting value x_0.

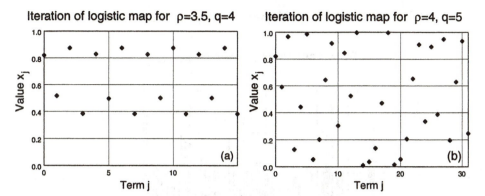

Figs. 10.2(a, b). Plot of x_j vs. j for iteration of the logistic map for (a) $\rho = 3.5$ with $(x_0, q) = (0.82, 4)$; and (b) $\rho = 4$ with $(x_0, q) = (0.82, 5)$.

Let us next investigate the power spectrum. Figure 10.3 shows the quantity $|c_k|^2$ as a function of the frequency $\nu = k/(n+1)$ as calculated from Eq. (10.14) for the two cases in Figs. 10.1(b) and 10.2(a).

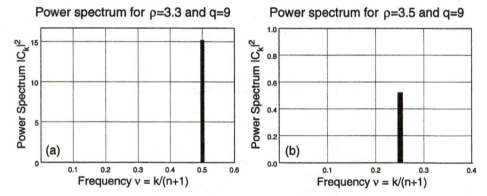

Figs. 10.3(a, b). Plot of $|c_k|^2$ vs. $\nu = k/(n+1)$ for the two cases: (a) as shown in Fig. 10.1(b), with $\rho = 3.3$ and $(x_0, q) = (0.82, 9)$; and (b), as shown in Fig. 10.2(a) with $\rho = 3.5$ and $(x_0, q) = (0.82, 9)$.

In Fig. 10.3(a), essentially all the strength lies at the period $\tau = 2$ (frequency $\nu = 1/2$) corresponding to the two-cycle oscillation seen in Fig. 10.1(b). In Fig. 10.3(b), the period doubling to $\tau = 4$ (frequency $\nu = 1/4$) observed with the four-cycle oscillation in Fig. 10.2(a) is identified — although this takes a little more examination, as the remaining dominant contribution at $\nu = 1/2$ has been suppressed in preparing this figure (note that the horizontal scale ends at 0.4).

Figure 10.4 shows the power spectrum for the chaotic case in Fig. 10.2(b). Within the fluctuations, which depend sensitively on the initial conditions and number of iteration points $n + 1 = 2^q$, the quantity $|c_k|^2$ is an essentially flat function of frequency $k/(n+1)$. No underlying periodicity is evident.

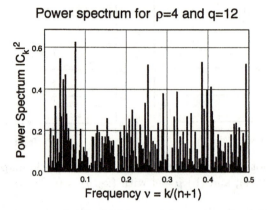

Fig. 10.4. Plot of $|c_k|^2$ vs. $\nu = k/(n+1)$ for the case in Fig. 10.2(b), with $\rho = 4$ and $(x_0, q) = (0.82, 12)$. Only every tenth point is plotted here.

Extended numerical analysis of this type leads to the "phase diagram" for the logistic map shown in Fig. 10.5.

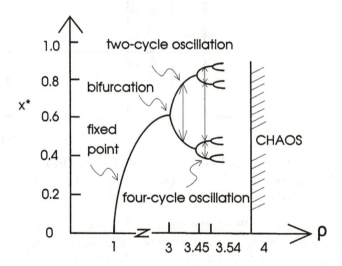

Fig. 10.5. Sketch of stability lines x^* as a function of the parameter ρ in the logistic map as determined from numerical analysis. Note the expanded scale. For a numerical plot of this diagram with a linear scale, see Fig. 7.29 of [Jo98].

The plot shows the stable values of x as a function of ρ. These are:

- For values of $\rho < 1$, the iteration converges to a value $x^* = 0$;

92

- For $1 < \rho < 3$, the iteration converges to a value $x^\star$ given by

$$x^\star \; = \; 1 - \frac{1}{\rho} \qquad\qquad ; 1 < \rho < 3 \qquad\qquad (10.16)$$

This behavior is illustrated in Fig. 10.1(a);

- At $\rho = 3$, there is a *bifurcation*;

- For $3 < \rho < 3.449 \cdots$, there is a two-cycle oscillation with period $\tau = 2$ (frequency $\nu = 1/2$) between the indicated values of $x^\star_{\min}$ and $x^\star_{\max}$ [see Fig. 10.1(b)];

- At $\rho \approx 3.449 \cdots$, there is a second pair of bifurcations;

- For $3.449 < \rho < 3.54$, there is a four-cycle oscillation with period $\tau = 4$ (frequency $\nu = 1/4$) between the values $(x^\star_1, \cdots, x^\star_4)$ shown in the figure [see Fig. 10.2(a)];

- This pattern keeps repeating [Fe78, Fe80]. There is a sequence of values ρ_n for which the 2^n cycle becomes stable, with a limiting value of $\rho_\infty = 3.5699 \cdots$;

- Above this value, one observes irregular chaotic behavior, but in certain ranges, odd regularities are found. For example, at $\rho \approx 3.81$, a three-cycle periodic structure appears. Upon blow-up of small regions, a self-similar, or *fractal*, structure is observed. The whole picture is very complex and fascinating;[54]

- At $\rho = 4$, *chaos* is clearly evident [see Fig. 10.2(b)].

Can we understand some of these phenomena in more detail? Let us investigate the problem using analytical methods, which turn out to give a valuable perspective.

Some analytic results

Kadanoff has written a very nice review article, intended for a general physics audience; when combined with our previous discussion, it helps one to understand much of this behavior [Ka83].

Let us look for *fixed points* of the recursion relation. These would be values $x^\star$ that are reproduced upon iteration so that

$$x^\star \; = \; \rho\, x^\star (1 - x^\star) \qquad\qquad (10.17)$$

This equation has two solutions

$$x^\star \; = \; 0$$
$$x^\star \; = \; 1 - \frac{1}{\rho} \qquad\qquad (10.18)$$

---

[54]For a detailed account of these phenomena, see, for example, [Jo98b, Ot02b].

The first solution indicates a fixed point at the origin. Consider the linearized theory near the origin. The linearized form of the recursion relation in Eq. (10.9), as well as its immediate solution, are

$$
\begin{aligned}
x_{j+1} &= \rho x_j \\
x_j &= \rho^j x_0
\end{aligned}
\tag{10.19}
$$

This solution evidently goes to zero as $j \to \infty$ provided $\rho < 1$. In this range of ρ, the origin is a *stable fixed point*. If $\rho = 1$, this linearized solution remains constant near x_0 as $j \to \infty$, and the linearized solution diverges if $\rho > 1$, indicating that the fixed point then becomes unstable.

Consider the second fixed point in Eq. (10.18) with $\rho > 1$. Let us linearize about this value, writing $x_j = x^* + \delta x_j$

$$
\begin{aligned}
\delta x_{j+1} &= \rho \left(\frac{1}{\rho} - 1 + \frac{1}{\rho} \right) \delta x_j \\
&= (2 - \rho)\, \delta x_j
\end{aligned}
\tag{10.20}
$$

This implies that any variation will iterate to zero provided[55]

$$
\begin{aligned}
|2 - \rho| &< 1 \\
1 < \rho &< 3
\end{aligned}
\tag{10.21}
$$

The second fixed point is *stable* for values of ρ lying in this range. For values of $\rho > 3$, the second fixed point is clearly unstable.

The above fixed point is one where the value of x is reproduced upon a single iteration. Suppose instead that the value of x repeats only after *two* iterations. Define

$$
F_\rho(x) = \rho x (1 - x)
\tag{10.22}
$$

The logistic map is then of the form

$$
x_{j+1} = F_\rho(x_j)
\tag{10.23}
$$

and the condition for the previous fixed point in Eq. (10.17) is

$$
x^* = F_\rho(x^*)
\tag{10.24}
$$

For the new type of fixed point, we therefore require

$$
x^* = F^{(2)}(x^*) = F_\rho[\, F_\rho(x^*)\,]
\tag{10.25}
$$

or, in the present case

$$
x^* = \rho \left[\rho x^*(1 - x^*) \right] \left[1 - \rho x^*(1 - x^*) \right]
\tag{10.26}
$$

---

[55]Evidently the variation will oscillate to zero in the range $2 < \rho < 3$.

This relation is a quartic equation for x^*. Two of the roots will be our previous solutions in Eq. (10.18), since by Eq. (10.24) they obviously satisfy Eq. (10.25); it turns out, however, that they now represent unstable fixed points. These roots can be factored from the above quartic equation, which can then be written in factored form as

$$x^* \left[x^* - \left(1 - \frac{1}{\rho} \right) \right] \left[\rho^3 x^{*2} - \rho^2 (1 + \rho) x^* + \rho(1 + \rho) \right] = 0 \quad (10.27)$$

The remaining quadratic equation has the roots[56]

$$x_\pm = \frac{1}{2\rho}(1 + \rho) \pm \frac{1}{2\rho} \sqrt{(\rho - 3)(\rho + 1)} \quad (10.28)$$

At $\rho = 3$, one has $x_\pm = 2/3 = 1 - 1/\rho$ reproducing the previous fixed point. For $\rho > 3$, in contrast, the two roots $x_\pm$ are both *real and distinct*. It is readily established that these new roots $x_\pm$ satisfy[57]

$$\begin{aligned} F_\rho(x_+) &= x_- \\ F_\rho(x_-) &= x_+ \end{aligned} \quad (10.29)$$

This result explains the period $\tau = 2$ oscillation. Such an analysis locates the values of $x^*_{max} = x_+$ and $x^*_{min} = x_-$ in Fig. 10.5. A linear stability analysis about these new fixed points, which we leave as a problem, then determines that this period-two oscillation is stable for $3 < \rho < 1 + \sqrt{6} = 3.4494 \cdots$. The analysis can be extended to the higher cycles and clearly becomes more and more intricate, but it is conceptually well defined.

Finally, consider the behavior for $\rho = 4$, which we have already seen to be *chaotic*.[58] For this special value of ρ, change variables in the recursion relation and define

$$x_j = \frac{1}{2}(1 - \cos 2\pi\theta_j) \quad (10.30)$$

With the aid of the trigonometric identity $\cos 2\chi = 1 - 2\sin^2 \chi$, the recursion

---

[56]The "star" is implied on these roots; it is omitted for notational simplicity.
[57]These relations must hold, for let

$$F_\rho(x_+) = x_g$$

where x_g is some value of x other than those previously found in Eqs. (10.18). Upon insertion of this expression in Eq. (10.23), one has

$$F_\rho(x_g) = F_\rho[F_\rho(x_+)] = x_+$$

and one more iteration gives

$$F_\rho[F_\rho(x_g)] = F_\rho(x_+) = x_g$$

Hence x_g is a new fixed point that is neither x_+ nor a previous root; hence, it must be x_-.
[58]Note that the maximum value of $x(1 - x)$ in the interval $[0, 1]$ is $1/4$; thus $\rho = 4$ is the largest value of the parameter ρ for which the iterated x is restricted to this interval.

relation Eq. (10.9) becomes

$$\frac{1}{2}(1 - \cos 2\pi\theta_{j+1}) = 4\frac{1}{2}(1 - \cos 2\pi\theta_j)\frac{1}{2}(1 + \cos 2\pi\theta_j)$$

$$= \sin^2 2\pi\theta_j$$

$$= \frac{1}{2}(1 - \cos 4\pi\theta_j) \tag{10.31}$$

The solution to this new recursion relation is given by

$$\theta_{j+1} = 2\theta_j$$
$$\theta_j = 2^j\,\theta_0 \tag{10.32}$$

where the second line follows exactly as in Eq. (10.19).

One observes from Eq. (10.30) that either of the two following replacements yields exactly the same x_j

$$\theta_j \;\rightarrow\; \theta_j + \text{integer}$$
$$\theta_j \;\rightarrow\; -\theta_j \tag{10.33}$$

Now any real number θ_j has the following structure

$$\theta_j = XXX \cdots XXX.yyyy \cdots \tag{10.34}$$

The integer part of this number "$XXX \cdots XXX$" is irrelevant since it can be transformed away by the first of Eqs. (10.33)! Thus the answer depends only on the decimal ".$yyyy \cdots$" part of θ_j. Hence the answer *is extremely sensitive to the initial conditions and round-off error.*

The translation and reflection invariances in Eqs. (10.33) imply that any value of θ_j can be mapped into the interval

$$0 \le \;\theta_j\; \le \frac{1}{2} \tag{10.35}$$

We can thus confine our considerations to this region of θ_j. The mappings of Eq. (10.32) that then keep us in this interval are

$$\theta_{j+1} = 2\theta_j \qquad\qquad ; 0 \le \theta_j \le \frac{1}{4}$$
$$\theta_{j+1} = -(2\theta_j - 1)$$
$$= 1 - 2\theta_j \qquad ; \frac{1}{4} \le \theta_j \le \frac{1}{2} \tag{10.36}$$

Now consider the resulting sequences θ_j generated with these relations starting from a representative set of initial values

$$\theta_0 = \frac{1}{3} \qquad ; \frac{1}{3}, \frac{1}{3}, \frac{1}{3}, \cdots$$
$$\theta_0 = \frac{1}{5} \qquad ; \frac{1}{5}, \frac{2}{5}, \frac{1}{5}, \frac{2}{5}, \cdots$$
$$\theta_0 = \frac{1}{9} \qquad ; \frac{1}{9}, \frac{2}{9}, \frac{4}{9}, \frac{1}{9}, \frac{2}{9}, \frac{4}{9}, \cdots \tag{10.37}$$

One observes:

- The first sequence yields a fixed point;

- The second sequence is a periodic two-cycle;

- The third sequence is a periodic three-cycle;

- Similar observations show that the chaotic region $\rho = 4$ contains *sequences with all possible periods*.

This is the result that we found empirically in Fig. 10.4.

11 Liouville's theorem revisited

Dynamical systems with a finite number of degrees of freedom can follow various scenarios when they become increasingly chaotic. We have illustrated some of them with the Lorenz equations (three coupled, first-order, nonlinear, ordinary differential equations) and with the logistic map (a first-order, nonlinear, finite-difference equation). To provide a different and very important example, we now return to hamiltonian dynamics from Sec. 8.

It is a remarkable formal result that the dynamical evolution of a hamiltonian system is itself a canonical transformation. To demonstrate this assertion, consider first a special form of the generator of the transformation in Eqs. (35.4)

$$S_0(q, P) \;=\; \sum_\sigma q_\sigma P_\sigma \tag{11.1}$$

Here $\sigma = 1, \cdots, n$ runs over all the generalized coordinates. It follows from Eqs. (35.4) that

$$p_\sigma \;=\; \frac{\partial S_0(q, P)}{\partial q_\sigma} \;=\; P_\sigma$$

$$Q_\sigma \;=\; \frac{\partial S_0(q, P)}{\partial P_\sigma} \;=\; q_\sigma \tag{11.2}$$

Thus S_0 is the generator of the *identity* transformation.

Next take the following modified generator

$$
\begin{aligned}
S(q, P, t) \;&=\; S_0(q, P) + H\,dt \\
&=\; \sum_\sigma q_\sigma P_\sigma + H(p, q, t)\,dt
\end{aligned}
\tag{11.3}
$$

which differs from the generator S_0 of the identity transformation by the infinitesimal quantity $H\,dt$. It follows from Eqs. (35.4) that

$$
\begin{aligned}
p_\sigma \;&=\; \frac{\partial S(q, P, t)}{\partial q_\sigma} \;=\; P_\sigma + \left[\frac{\partial H(p, q, t)}{\partial q_\sigma}\right]_{P_\sigma} dt \\
&=\; P_\sigma + \frac{\partial H(p, q, t)}{\partial q_\sigma}\,dt + \cdots \\
&=\; P_\sigma - \frac{dp_\sigma}{dt}\,dt + \cdots
\end{aligned}
\tag{11.4}
$$

The first line shows that P_σ differs from p_σ by contributions of order dt. Hence one can replace $P_\sigma \approx p_\sigma$ in the last term since it already contains the explicit factor dt. Hamilton's Eq. (32.30) has then been invoked in arriving at the final result. Thus, to first order in dt

$$P_\sigma = p_\sigma + \frac{dp_\sigma}{dt}dt$$
$$P_\sigma = p_\sigma(t + dt) \qquad ; \text{ to first order} \qquad (11.5)$$

To this order, the new momentum is simply the old momentum displaced forward in time from t to $t + dt$, which follows from Taylor's theorem. Similarly,

$$Q_\sigma = \frac{\partial S}{\partial P_\sigma} = q_\sigma + \frac{\partial H}{\partial P_\sigma}dt \qquad (11.6)$$

and in like fashion

$$Q_\sigma = q_\sigma + \frac{\partial H(p,q,t)}{\partial p_\sigma}dt + \cdots$$
$$= q_\sigma + \frac{dq_\sigma}{dt}dt + \cdots$$
$$Q_\sigma = q_\sigma(t + dt) \qquad ; \text{ to first order} \qquad (11.7)$$

Here the other Hamilton's Eq. (32.29) has been used in the second line. The new canonical coordinates (P_σ, Q_σ) thus are simply the dynamical evolutions of the original canonical variables (p_σ, q_σ), to first order in dt. In a well-defined sense, the hamiltonian H in Eq (11.3) is the generator of this infinitesimal time translation.

This result has one central implication. Consider the $2n$-dimensional phase space (p_σ, q_σ), $\sigma = 1, \cdots, n$. Start at some point $[p_\sigma(0), q_\sigma(0)]$ at $t = 0$. As the system evolves, this point traces out a path in phase space. Consider a small volume element in phase space at the initial point

$$dV = dp_1 \cdots dp_n dq_1 \cdots dq_n \qquad (11.8)$$

or, if we take a small finite volume element

$$\text{Vol} = \int dV = \int dp_1 \cdots dp_n dq_1 \cdots dq_n \qquad (11.9)$$

A remarkable theorem, due to Liouville, states that this volume is *invariant* along the dynamical trajectory. Although its shape may change, the volume itself does not.[59] We give a proof following [Wa89].

After an infinitesimal time interval dt, the multidimensional volume element has changed to

$$dV' = dP_1 \cdots dP_n dQ_1 \cdots dQ_n$$
$$= \frac{\partial(P,Q)}{\partial(p,q)}dp_1 \cdots dp_n dq_1 \cdots dq_n \qquad (11.10)$$

---

[59]The theorem is actually more general; the volume is invariant under any canonical transformation, although we do not prove this result here. For a concise and elegant general proof that relies on basic properties of jacobian determinants, see [La60].

where the factor in front is the $2n \times 2n$ jacobian determinant for the transformation

$$\frac{\partial(P,Q)}{\partial(p,q)} = \begin{vmatrix} \partial P_1/\partial p_1 & \partial Q_1/\partial p_1 & \cdots & \partial P_n/\partial p_1 & \partial Q_n/\partial p_1 \\ \partial P_1/\partial q_1 & \partial Q_1/\partial q_1 & \cdots & \partial P_n/\partial q_1 & \partial Q_n/\partial q_1 \\ \vdots & \vdots & \vdots & \vdots & \vdots \\ \partial P_1/\partial p_n & \partial Q_1/\partial p_n & \cdots & \partial P_n/\partial p_n & \partial Q_n/\partial p_n \\ \partial P_1/\partial q_n & \partial Q_1/\partial q_n & \cdots & \partial P_n/\partial q_n & \partial Q_n/\partial q_n \end{vmatrix} \quad (11.11)$$

The elements in this determinant are readily evaluated to order dt from the transformation in Eqs. (11.4) and (11.7)

$$\frac{\partial P_\sigma}{\partial p_\rho} = \delta_{\sigma\rho} - \frac{\partial^2 H}{\partial p_\rho \partial q_\sigma} dt \quad ; \quad \frac{\partial P_\sigma}{\partial q_\rho} = -\frac{\partial^2 H}{\partial q_\rho \partial q_\sigma} dt$$

$$\frac{\partial Q_\sigma}{\partial q_\rho} = \delta_{\sigma\rho} + \frac{\partial^2 H}{\partial q_\rho \partial p_\sigma} dt \quad ; \quad \frac{\partial Q_\sigma}{\partial p_\rho} = \frac{\partial^2 H}{\partial p_\rho \partial p_\sigma} dt \quad (11.12)$$

In evaluating the determinant in Eq. (11.11), each term containing a nondiagonal element as a factor contains at least one other nondiagonal element as a factor. Since all these elements are proportional to dt, these terms are therefore at least of second order in dt and do not contribute to first order. To this order, there remains then only the contribution from the diagonal elements

$$\frac{\partial(P,Q)}{\partial(p,q)} = \prod_\sigma \left(1 - \frac{\partial^2 H}{\partial p_\sigma \partial q_\sigma} dt\right)\left(1 + \frac{\partial^2 H}{\partial q_\sigma \partial p_\sigma} dt\right) + O(dt^2)$$

$$= 1 + \sum_\sigma \left(\frac{\partial^2 H}{\partial p_\sigma \partial q_\sigma} - \frac{\partial^2 H}{\partial p_\sigma \partial q_\sigma}\right) dt + O(dt^2)$$

$$= 1 + O(dt^2) \quad (11.13)$$

where the order of the partial derivatives has been interchanged in the second line, the two terms then canceling. Thus the jacobian determinant is one, and the theorem holds for infinitesimal time displacements. It remains to extend it to finite times, which can be done as follows.

Write the time development of the volume element as

$$dV' = J(t,t')dV \quad (11.14)$$

with the initial condition

$$J(t,t) = 1 \quad (11.15)$$

For infinitesimal time displacements, we have just shown that

$$J(t,t+dt) = J(t,t) + \left.\frac{\partial J(t'',t)}{\partial t}\right|_{t''=t} dt$$

$$= 1 + \left.\frac{\partial J(t'',t)}{\partial t}\right|_{t''=t} dt$$

$$= 1 \quad (11.16)$$

Thus

$$\frac{\partial J(t'',t)}{\partial t}\bigg|_{t''=t} = 0 \qquad (11.17)$$

for any t.

Now in Eq. (11.14), consider an arbitrary third time t'' lying between t and t'. Then going from t to t'' one has

$$dV'' = J(t,t'')dV \qquad (11.18)$$

and from t'' to t'

$$dV' = J(t'',t')dV'' \qquad (11.19)$$

The overall effect is

$$\begin{aligned}
dV' &= J(t'',t')J(t,t'')dV \\
&= J(t,t'')J(t'',t')dV \\
&= J(t,t')dV \qquad (11.20)
\end{aligned}$$

where the second line follows since J is just a scalar function. Thus

$$J(t,t') = J(t,t'')J(t'',t') \qquad (11.21)$$

because $J(t,t'')J(t'',t')$ must be the same as $J(t,t')$ for any t'' between t and t'. Differentiate this relation with respect to the final time t', and then set $t'' = t'$

$$\frac{\partial J(t,t')}{\partial t'} = J(t,t')\frac{\partial J(t'',t')}{\partial t'}\bigg|_{t''=t'}$$

$$\frac{\partial J(t,t')}{\partial t'} = 0 \qquad (11.22)$$

where Eq. (11.17) has been invoked in the final result. This relation holds for arbitrary t' and provides a first-order differential equation for the jacobian determinant. For $t' = t$, one has the initial condition $J(t,t) = 1$ of Eq. (11.15), so that

$$J(t,t') = 1 \qquad ; \text{ arbitrary } t' \qquad (11.23)$$

Hence from Eq. (11.14)

$$dV' = dV \qquad ; \text{ arbitrary } t' \qquad (11.24)$$

which is the result to be proven.

For *any* hamiltonian system, one concludes that the magnitude of the volume element in phase space does not change in the course of its motion along a phase orbit. This is Liouville's theorem.[60]

---

[60]The reader may indeed feel that we have spent an inordinate amount of time on Liouville's theorem. As justification, we recall that this result is one of the cornerstones of classical statistical mechanics, so that a clear understanding is important [Wa89]. Note that the Lorenz equations, which are *not* a hamiltonian system, illustrate another possible situation, namely, the continuous shrinking of the phase volume seen in Eq. (9.24).

12   Action-angle variables revisited

The subsequent analysis in this book will be based on action-angle variables, as employed in Sec. 8. We start the discussion with a rather complete review of the action-angle variables for a many-body collection of n separable, periodic systems from Sec. 36. In the following Sec. 13, we consider periodic, one-body perturbations of these systems and then in Sec. 14, we introduce couplings among these originally independent degrees of freedom.

Separable, periodic hamiltonian systems

Consider n generalized coordinates and a time-independent hamiltonian. In this case, the Hamilton-Jacobi generating function S has a simpler form given by Eq. (35.25) and obeys the associated partial differential equation given by Eqs. (35.26)

$$S(q_1, \cdots, q_n, \alpha_1, \cdots, \alpha_n, t) = W(q_1, \cdots, q_n, \alpha_1, \cdots, \alpha_n) - \alpha_1 t$$

$$H\left(\frac{\partial W}{\partial q_1}, \cdots, \frac{\partial W}{\partial q_n}, q_1, \cdots, q_n\right) = \alpha_1 = E \qquad (12.1)$$

Here $(\alpha_1, \alpha_2, \cdots, \alpha_n)$ are a set of n independent nonadditive integration constants for the Hamilton-Jacobi equation, with $\alpha_1 \equiv E$. In the special case of n *separable*, integrable systems, the Hamilton-Jacobi characteristic function W takes the form in Eqs. (35.27) and (36.1)

$$W(q_1, \cdots, q_n, \alpha_1, \cdots, \alpha_n) = W_1(q_1, \alpha_1, \cdots, \alpha_n) + \cdots +$$
$$W_n(q_n, \alpha_1, \cdots, \alpha_n) \qquad (12.2)$$

The corresponding momenta are expressed in terms of this separated generating function by Eqs. (35.14) and (35.19)

$$
\begin{aligned}
p_\sigma &= \frac{\partial S(q_1, \cdots, q_n, \alpha_1, \cdots, \alpha_n, t)}{\partial q_\sigma} \\[2mm]
&= \frac{\partial W(q_1, \cdots, q_n, \alpha_1, \cdots, \alpha_n)}{\partial q_\sigma} \\[2mm]
&= \frac{\partial W_\sigma(q_\sigma, \alpha_1, \cdots, \alpha_n)}{\partial q_\sigma} \qquad ; \sigma = 1, \cdots, n \qquad (12.3)
\end{aligned}
$$

We assume that each of the n separable, integrable systems undergoes periodic motion, with (in general) different periods. The action for the σth system is then given by Eq. (36.2)[61]

$$
\begin{aligned}
J_\sigma &= \frac{1}{2\pi} \oint p_\sigma \, dq_\sigma \\[2mm]
&= \frac{1}{2\pi} \oint \frac{\partial W_\sigma(q_\sigma, \alpha_1, \cdots, \alpha_n)}{\partial q_\sigma} dq_\sigma \\[2mm]
&= J_\sigma(\alpha_1, \cdots, \alpha_n) \qquad\qquad ; \sigma = 1, \cdots, n \qquad (12.4)
\end{aligned}
$$

---

[61] Recall that we now include a factor of $1/2\pi$ in this definition.

It is assumed that these last relations are *invertible* so that one can express

$$\alpha_\sigma = \alpha_\sigma(J_1, \cdots, J_n) \qquad ; \sigma = 1, \cdots, n \qquad (12.5)$$

As in Sec. 8, we then use these relations to introduce a new set of constants in the Hamilton-Jacobi generating function[62]

$$
\begin{aligned}
\bar{S}(q_1, \cdots, q_n, J_1, \cdots, J_n, t) &= \bar{W}(q_1, \cdots, q_n, J_1, \cdots, J_n) - E(J_1, \cdots, J_n)t \\
&= \bar{W}_1(q_1, J_1, \cdots, J_n) + \cdots + \bar{W}_n(q_n, J_1, \cdots, J_n) \\
&\quad - E(J_1, \cdots, J_n)t \qquad (12.6)
\end{aligned}
$$

From Hamilton-Jacobi theory it follows that [see Eqs. (36.10) and (36.12)]

$$
\begin{aligned}
\bar{\beta}_\sigma &= \frac{\partial \bar{S}(q_1, \cdots, q_n, J_1, \cdots, J_n, t)}{\partial J_\sigma} \\
&= \frac{\partial \bar{W}(q_1, \cdots, q_n, J_1, \cdots J_n)}{\partial J_\sigma} - \frac{\partial E(J_1, \cdots, J_n)}{\partial J_\sigma} t \\
&= \text{constant} \qquad ; \sigma = 1, \cdots, n \qquad (12.7)
\end{aligned}
$$

The angle variables (now denoted ϕ_σ) are defined by

$$\phi_\sigma = \frac{\partial \bar{W}(q_1, \cdots, q_n, J_1, \cdots J_n)}{\partial J_\sigma} \qquad (12.8)$$

It follows from Eq. (12.7) that they satisfy

$$
\begin{aligned}
\phi_\sigma &= \omega_\sigma t + \bar{\beta}_\sigma \\
\omega_\sigma &= \frac{\partial E(J_1, \cdots, J_n)}{\partial J_\sigma} \qquad ; \sigma = 1, \cdots, n \qquad (12.9)
\end{aligned}
$$

The angle variables all increase linearly with the time, and each system is periodic in the relevant ϕ_σ.

Now assume that after a (in general long) period of time Δt, all the individual systems have executed some integral number of periods, which implies that the frequencies are commensurate. This is an inessential assumption, since if it were not true, the physical parameters in the problem could be varied in a completely negligible manner until the ratio of all the frequencies is indeed rational.[63] What is the corresponding change in the angle variables over the time period Δt? Since the $\{J_\sigma\}$ are constants of the motion, the angle variables in Eq. (12.8) change only because the coordinates $\{q_\sigma\}$ change, which is described by the differential relation

$$
\begin{aligned}
d\phi_\sigma &= \sum_\lambda \frac{\partial \phi_\sigma}{\partial q_\lambda} dq_\lambda \\
&= \sum_\lambda \frac{\partial^2 \bar{W}(q_1, \cdots, q_n, J_1, \cdots, J_n)}{\partial q_\lambda \partial J_\sigma} dq_\lambda \qquad (12.10)
\end{aligned}
$$

---

[62]It is really $\bar{E}(J_1, \cdots, J_n)$, but we omit the bar here as superfluous.
[63]We make the assumption because we believe that it clarifies the presentation.

where Eq. (12.8) has been introduced in the second line. A change in the order of partial derivatives, and the use of Eq. (12.6) reduces this equation to

$$d\phi_\sigma = \sum_\lambda \frac{\partial^2 \bar{W}(q_1,\cdots,q_n\, J_1,\cdots,J_n)}{\partial J_\sigma \partial q_\lambda} dq_\lambda$$

$$= \sum_\lambda \frac{\partial^2 \bar{W}_\lambda(q_\lambda,\, J_1,\cdots,J_n)}{\partial J_\sigma \partial q_\lambda} dq_\lambda \qquad (12.11)$$

With the identification of p_λ from Eq. (12.3), this relation becomes

$$d\phi_\sigma = \frac{\partial}{\partial J_\sigma} \sum_\lambda p_\lambda\, dq_\lambda \qquad (12.12)$$

Consider the entire time period Δt, during which each degree of freedom q_σ experiences an integral number n_σ of periods τ_σ, so that[64]

$$\Delta t = n_\sigma \tau_\sigma \qquad\qquad ;\, \sigma = 1,\cdots,n \qquad (12.13)$$

The total change in the angle variable ϕ_σ, which is linear in the time, is

$$\Delta \phi_\sigma = \omega_\sigma n_\sigma \tau_\sigma \qquad\qquad ;\, \sigma = 1,\cdots,n \qquad (12.14)$$

On the other hand, the total change in the same angle variable can be obtained by integrating Eq. (12.12) (recall that the $\{J_\sigma\}$ are constants of the motion)

$$\Delta \phi_\sigma = \frac{\partial}{\partial J_\sigma} \sum_\lambda \oint p_\lambda\, dq_\lambda$$

$$= \frac{\partial}{\partial J_\sigma} \sum_\lambda 2\pi n_\lambda J_\lambda$$

$$= 2\pi n_\sigma \qquad\qquad ;\, \sigma = 1,\cdots,n \qquad (12.15)$$

This result demonstrates that each system is 2π-periodic in its angle variable. A comparison of the two expressions for $\Delta \phi_\sigma$ in Eqs. (12.14) and (12.15) leads to the conclusion that

$$\omega_\sigma \tau_\sigma = 2\pi$$

$$\omega_\sigma = \frac{\partial E(J_1,\cdots,J_n)}{\partial J_\sigma} \qquad ;\, \sigma = 1,\cdots,n \qquad (12.16)$$

These relations express the fundamental time periods of each of the separable, periodic systems in terms of a partial derivative of the energy expressed in terms of the individual actions. We give two examples:

---

[64]Note that there is no summation convention on repeated indices used in this section.

Simple harmonic oscillators. Consider a collection of n simple harmonic oscillators, with individual angular frequencies $\omega_{0\sigma}$, $\sigma = 1, \cdots, n$, and hamiltonian

$$H = \sum_\sigma h(p,q)_\sigma$$

$$h(p,q) = \frac{p^2}{2m} + \frac{1}{2}m\omega_0^2 q^2 \qquad (12.17)$$

Here the notation $h(p,q)_\sigma$ implies that all quantities in the one-body hamiltonian are evaluated as appropriate for the σth system. Now assume a separated characteristic function of the form in Eq. (12.2). The Hamilton-Jacobi equation then becomes

$$\sum_\sigma h\left(\frac{dW}{dq}, q\right)_\sigma = E$$

$$h\left(\frac{dW}{dq}, q\right) = \frac{1}{2m}\left(\frac{dW}{dq}\right)^2 + \frac{1}{2}m\omega_0^2 q^2 \qquad (12.18)$$

The individual energies E_σ are conserved, and they can be taken as the *separation constants*, giving

$$\left[\frac{1}{2m}\left(\frac{dW}{dq}\right)^2 + \frac{1}{2}m\omega_0^2 q^2\right]_\sigma = E_\sigma$$

$$\sum_\sigma E_\sigma = E \qquad (12.19)$$

The problem has now been reduced to that of the one-body oscillator analyzed in Sec. 8, and from Eq. (8.34) one has for each individual oscillator

$$E_\sigma = (\omega_0 J)_\sigma \qquad (12.20)$$

Hence for the collection of oscillators

$$E(J_1, \cdots, J_n) = \sum_\sigma (\omega_0 J)_\sigma \qquad (12.21)$$

Thus from Eq. (12.9)

$$\omega_\sigma = \frac{\partial E(J_1, \cdots, J_n)}{\partial J_\sigma}$$

$$= \omega_{0\sigma} \qquad ; \sigma = 1, \cdots, n \qquad (12.22)$$

The angular frequencies of the angle variables in the many-body action-angle analysis are just the individual oscillator frequencies — not a very surprising result. The separated n-body characteristic function now has the form

$$\bar{W}(q_1, \cdots, q_n, J_1, \cdots, J_n) = \sum_\sigma \bar{W}_\sigma(q_\sigma, J_\sigma) \qquad (12.23)$$

104

The angle variables each develop in time according to Eq. (12.9)

$$\phi_\sigma \;=\; \omega_\sigma t + \bar{\beta}_\sigma \qquad\qquad ; \sigma = 1, \cdots, n \qquad (12.24)$$

Their relation to the coordinate q_σ is obtained from Eqs. (12.23) and (8.36)

$$
\begin{aligned}
\phi_\sigma &= \frac{\partial \bar{W}(q_1, \cdots, q_n, J_1, \cdots, J_n)}{\partial J_\sigma} \\[2mm]
&= \frac{\partial \bar{W}_\sigma(q_\sigma, J_\sigma)}{\partial J_\sigma} \qquad ; \sigma = 1, \cdots, n \qquad (12.25)
\end{aligned}
$$

and the one-body analysis in Sec. 8 then leads to the results in Eqs. (8.40)[65]

$$
\begin{aligned}
p_\sigma(t) &= \sqrt{(2m\omega_0 J)_\sigma}\, \cos\phi_\sigma(t) \\[2mm]
q_\sigma(t) &= \sqrt{\left(\frac{2J}{m\omega_0}\right)_\sigma}\, \sin\phi_\sigma(t) \\[2mm]
\phi_\sigma(t) &= \omega_{0\sigma}\, t + \bar{\beta}_\sigma \qquad ; \sigma = 1, \cdots, n \qquad (12.26)
\end{aligned}
$$

Pendulums. For a collection of pendulums, each undergoing small-amplitude libration, Eq. (8.66) tells us that the analysis is exactly the same as for the oscillators. Here the angular frequency of each individual pendulum is given by $\omega_{0\sigma} = \sqrt{g/l_\sigma}$. For librations of arbitrary amplitude, the more general one-body analysis of Sec. 8 applies to each pendulum.

If the n pendulums all undergo rotation in the high-energy limit,[66] then Eq. (8.56) implies that the only modification of the oscillator analysis is to replace Eqs. (12.21) and (12.22) by

$$
\begin{aligned}
E(J_1, \cdots, J_n) &= \sum_\sigma \left(\frac{J^2}{2ml^2}\right)_\sigma \\[2mm]
\omega_\sigma &= \frac{\partial E(J_1, \cdots, J_n)}{\partial J_\sigma} \\[2mm]
&= \left(\frac{J}{ml^2}\right)_\sigma = \dot{\theta}_\sigma \qquad ; \sigma = 1, \cdots, n \qquad (12.27)
\end{aligned}
$$

Each angular frequency ω_σ now depends explicitly on the corresponding action (and is precisely the associated angular velocity $\dot{\theta}_\sigma$). The angle variables are again given by Eq. (12.24) and (12.25), and the one-body analysis leads to the result in Eq. (8.69) that

$$\theta_\sigma(t) \;=\; \dot{\theta}_\sigma t + \theta_\sigma(0) \qquad\qquad ; \sigma = 1, \cdots, n \qquad (12.28)$$

for each coordinate in this high-energy limit. Results for rotations of any energy $E_\sigma > 2m_\sigma g l_\sigma$ follow as in Sec. 8.

---

[65]We restore an arbitrary origin and write $\phi = \omega t + \bar{\beta}$.

[66]In this limit, the physics very clear. For simplicity consider a single pendulum with $\dot{\theta} \gg \sqrt{g/l}$. Then $p = ml^2\dot{\theta}$ is large and constant, so that $J = (2\pi)^{-1}\oint p\, d\theta = p$. The energy then becomes $E = ml^2\dot{\theta}^2/2 = J^2/(2ml^2)$.

We note that although the action-angle variables (J_σ, ϕ_σ) are not the canon-ical pairs here, they differ from the canonical pairs $(J_\sigma, \bar{\beta}_\sigma)$ only by the time translations in Eqs. (12.9), and they do satisfy Hamilton's equations with the hamiltonian

$$\bar{H}(J_1, \cdots, J_n) \;=\; E(J_1, \cdots, J_n) \tag{12.29}$$

Thus

$$\frac{d\phi_\sigma}{dt} \;=\; \frac{\partial \bar{H}(J_1, \cdots, J_n)}{\partial J_\sigma} \;=\; \omega_{0\sigma}$$

$$\frac{dJ_\sigma}{dt} \;=\; -\frac{\partial \bar{H}(J_1, \cdots, J_n)}{\partial \phi_\sigma} \;=\; 0 \qquad ; \sigma = 1, \cdots, n \tag{12.30}$$

Phase plots and motion on tori

We next turn to the topic of the phase-space orbits of separable, periodic hamiltonian systems. We start with the simplest case of two such systems, and for definiteness consider two oscillators with hamiltonians and energies

$$h(p, q)_1 \;=\; \frac{1}{2m_1}p_1^2 + \frac{1}{2}m_1\omega_1^2 q_1^2 \;=\; E_1$$

$$h(p, q)_2 \;=\; \frac{1}{2m_2}p_2^2 + \frac{1}{2}m_2\omega_2^2 q_2^2 \;=\; E_2 \tag{12.31}$$

The oscillator frequencies are here simply denoted as (ω_1, ω_2). Phase space for this two-body system is four dimensional (q_1, p_1, q_2, p_2). Since the individual energies are conserved, there are two constraints, and thus the orbit actually lies on a two-dimensional surface in this four-dimensional space. To visualize this surface, start with the (q_1, p_1) plane, where particle 1 follows an elliptical trajectory in the clockwise direction (compare Fig. 1.1). There is a similar trajectory in the (q_2, p_2) plane for particle 2, and the combined motion is the union of these two orbits.

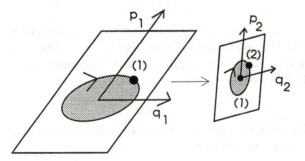

Fig. 12.1. Phase-space particle orbits: left—motion of particle (1) in the hor-izontal (q_1, p_1)plane (compare Fig. 1.1); right—motion of particle (2) in the (q_2, p_2) plane whose origin lies at the instantaneous position of particle (1) and which is oriented normal to the orbit of particle (1).

106

For definiteness, assume that $(J/m\omega)_1 > (J/m\omega)_2$ and $(Jm\omega)_1 > (Jm\omega)_2$, which means that both dimensions of orbit 2 are smaller than those of orbit 1 [see Eqs. (12.26)].

Now introduce the coordinate systems shown in Fig. 12.1, where the (q_2, p_2) plane is perpendicular to the (q_1, p_1) plane with the p_2 axis along the normal to the (q_1, p_1) plane. In addition, orient the normal to the (q_2, p_2) plane along the orbit of particle 1 and locate the origin of the (q_2, p_2) plane at the instantaneous position of particle 1. As time evolves, the phase-space ellipse for particle 2 sweeps out a torus or doughnut. The phase-space orbit of the two-particle system lies on this toroidal surface. This geometry is illustrated in Fig. 12.2.[67]

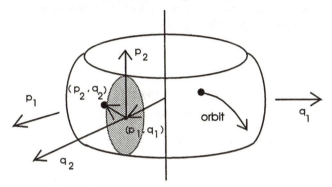

Fig. 12.2. Illustration of torus in (q_1, p_1, q_2, p_2) phase space for two separable periodic oscillator systems as a union of the phase space particle orbits in Fig. 12.1. It is assumed here that $(J/m\omega)_1 > (J/m\omega)_2$ and $(Jm\omega)_1 > (Jm\omega)_2$ [see Eqs. (12.26)].

The angular frequency for motion around the doughnut is ω_1, and the angular frequency for motion through the hole is ω_2. The combined trajectory is a helical spiral traced on the surface of this torus (one can think of a helix on a cylinder like a barber pole that is then wrapped around to form a ring). If the ratio ω_2/ω_1 is rational, the helical orbit winding around the torus will eventually close on itself (see later). If the ratio is irrational, the orbit will never close; nevertheless, it will eventually come arbitrarily close to any point on the torus.

---

[67]When viewed from above, the left hand side of Fig. 12.1 has the same right-handed (q_1, p_1) coordinate system and clockwise orbit as in Fig. 1.1. Define the normal to the plane as $\hat{n}_1 \equiv \hat{q}_1 \times \hat{p}_1$. The view from above is then opposite to the normal. When viewed from below the plane, (namely along the normal $\hat{n}_1$), the (q_1, p_1) coordinate system becomes left-handed and the orbit moves counterclockwise (try it). In forming the union of the orbits in Fig. 12.1, there are four possibilities: normal to the first plane up or down and normal to the second plane aligned or opposed to the trajectory of particle (1). We are, of course, free to define the (q_1, p_1, q_2, p_2) phase space in any manner we wish, as long as the coordinates are linearly independent. In preparing Fig. 12.2, we have chosen this union to present a phase-space drawing in a perspective that clarifies the coordinates and the subsequent toroidal phase-space orbit of the combined systems. Here the configuration in the left hand side of Fig. 12.1 is obtained by viewing the horizontal plane in Fig. 12.2 from below (note that as drawn, the normal $\hat{n}_1$ is down), and the right hand side is obtained by viewing the vertical plane from behind (opposite to the normal $\hat{n}_2 \equiv \hat{q}_2 \times \hat{p}_2$).

The time evolution of the orbit is given most simply in terms of the action-angle variables of Eqs. (8.40) and (12.26)

$$
\begin{aligned}
p_\sigma(t) &= \sqrt{(2m\omega J)_\sigma}\,\cos\phi_\sigma(t) \\
q_\sigma(t) &= \sqrt{\left(\frac{2J}{m\omega}\right)_\sigma}\,\sin\phi_\sigma(t) \\
\phi_\sigma(t) &= \omega_\sigma t + \bar{\beta}_\sigma \qquad\qquad ; \sigma = 1,2 \qquad (12.32)
\end{aligned}
$$

As the actions (J_1, J_2) change, so do the size and shape of the torus.

Suppose one treats this problem numerically. Hamilton's equations provide four coupled first-order differential equations

$$
\begin{aligned}
\dot{p}_1 &= -m_1\omega_1^2 q_1 \\
m_1\dot{q}_1 &= p_1 \\
\dot{p}_2 &= -m_2\omega_2^2 q_2 \\
m_2\dot{q}_2 &= p_2 \qquad\qquad (12.33)
\end{aligned}
$$

These equations can be recast in the form

$$
\dot{\mathbf{r}}(t) = \underline{M}\,\mathbf{r}(t) \qquad ; \mathbf{r} = \begin{bmatrix} p_1 \\ q_1 \\ p_2 \\ q_2 \end{bmatrix} \qquad (12.34)
$$

Here $\underline{M}$ is a matrix of physical constants, and $\mathbf{r}(t)$ is a four-component vector that has some trajectory in the four-dimensional phase space. Given the initial condition $\mathbf{r}(0)$, one can simply step forward in time and determine the orbit. Visualization in four dimensions is difficult. One way to view the result is to record the points where the orbit intersects a given two-dimensional surface (or plane). This is known as a *Poincaré surface of section*. For example, suppose that every time the vector $\mathbf{r}(t)$ comes within some small distance of the "surfaces" where $p_1 = 0$ and $p_2 = 0$, one records the remaining two coordinates (q_1, q_2). This provides a series of intersection points in the (q_1, q_2) plane. It is easy to see what the result will be from Fig. 12.2. The condition $p_2 = 0$ confines one to the (q_1, p_1) plane slicing the torus, and then $p_1 = 0$ leaves one on the q_1 axis. If q_{01} and q_{02} denote the maximum coordinate displacement of the oscillators, the orbit will intersect the q_1 axis only at the points $(q_{01} \pm q_{02})$ and $(-q_{01} \pm q_{02})$. Thus the result of the numerical calculation is to fill in these four corners of a rectangle, leaving the rest of the rectangle empty. It is easy to understand this result physically, since the condition $(p_1 = 0, p_2 = 0)$ means each oscillator is at its turning point, and there are just four possibilities.

If the components of the four-dimensional vector $\mathbf{r}(t)$ are plotted in the coordinate system of Fig. 12.2, the orbit actually lies on the two-dimensional surface of the indicated three-dimensional torus. One can also compute Poincaré surfaces of section here. For example, suppose one just records those points within some small distance of $p_2 = 0$, so that the orbit intersects the (q_1, p_1) plane slicing the torus. Where will the points lie? The intersection of the orbit with that plane will evidently lie on two ellipses (just the inner and outer curves obtained with a horizontal slice through the middle of the doughnut normal to its symmetry axis); midway between these two ellipses is the physical orbit (q_1, p_1) for the first oscillator. Again this result is evident physically, since in this case one simply samples the orbit of the first oscillator when the second one has its maximum positive or negative displacement.

Clearly, these ideas can be extended to calculations of more complicated problems in higher numbers of dimensions. One can conceptually speak of nested tori with different actions in n dimensions (as we shall do). It is always possible to generate Poincaré surfaces of section numerically for the intersection of an orbit with a given plane. If the problem is complicated enough, however, the interpretation of the surfaces of section may parallel the story of the blind men and the elephant (namely the description looks very different to observers in different regions of phase space).

A simpler, and more direct, analysis of the motion for the problem at hand is obtained in the four-dimensional action-angle phase space $(J_1, \tilde{\phi}_1, J_2, \tilde{\phi}_2)$ where

$$\begin{aligned}
\tilde{\phi}_1 &= \phi_1 - \bar{\beta}_1 &= \omega_1 t \\
\tilde{\phi}_2 &= \phi_2 - \bar{\beta}_2 &= \omega_2 t
\end{aligned} \tag{12.35}$$

Both (J_1, J_2) are constants of the motion, and the orbit lies in the $(\tilde{\phi}_1, \tilde{\phi}_2)$ plane. Furthermore, since the orbit is periodic with period 2π for each angle variable,[68] all values of $(\tilde{\phi}_1, \tilde{\phi}_2)$ can be mapped into the interval $[0, 2\pi]$ for both angle variables. Since the motion is periodic in these variables, the values at each pair of parallel boundaries must coincide. The surface of the torus has been mapped into a square in the $(\tilde{\phi}_1, \tilde{\phi}_2)$ plane with opposite sides identified, which is topologically equivalent. The time evolution of the orbit in this space is given by Eqs. (12.35). When the orbit leaves one edge of the region, the 2π translations imply that it returns to the region at the opposite edge with the same value.[69] The situation is illustrated in Fig. 12.3. Again, if the ratio ω_2/ω_1 is commensurate, the orbit eventually closes on itself. If not, it comes arbitrarily close to any given point.

A slightly different but equivalent picture is to imagine the whole $(\tilde{\phi}_1, \tilde{\phi}_2)$ plane covered with a square grid that is the periodic extension of the square in Fig. 12.3. In this extended domain, the orbit is simply a straight line with slope ω_2/ω_1 that starts at the origin $(\tilde{\phi}_1 = 0, \tilde{\phi}_2 = 0)$. As this continuous orbit crosses a boundary between adjacent squares, it executes the motion illustrated by the discontinuous orbit in Fig. 12.3. This extended description makes especially clear the distinction between rational and irrational values of the ratio ω_2/ω_1. If the ratio is rational, the straight-line trajectory eventually crosses another grid point, and the orbit then exactly repeats the same motion. If the ratio is

---

[68]Note that the constant shift by $\bar{\beta}$ does not affect the periodicity.

[69]This motion is identical with that found in the computer game Pacman that was played on a TV screen.

irrational, the orbit will never exactly cross another grid point, but it will clearly come very close to some of them. Such a situation is said to be "quasiperiodic".

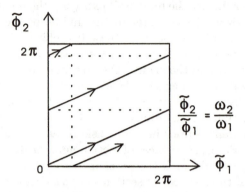

Fig. 12.3. Angle phase space $(\tilde{\phi}_1, \tilde{\phi}_2)$ for two separable periodic systems with given (J_1, J_2). The trajectory is given by Eq. (12.35). The orbit is periodic with period 2π in both the horizontal and vertical directions. Here $\tilde{\phi} = \phi - \bar{\beta}$, and it is assumed that $\omega_2/\omega_1 < 1$.

While our illustration has been for two oscillators, the action-angle phase space has the additional great advantage that it also holds for systems such as pendulums that exhibit both librational and rotational types of periodic motion. Despite the significantly more complicated dynamical motion in the original canonical (q, p) variables, these systems still have the *same straight-line angle orbits* shown in Fig. 12.3.

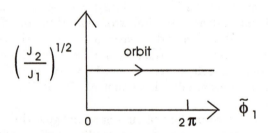

Fig. 12.4. Action-angle phase space $(\sqrt{J_2/J_1}, \tilde{\phi}_1)$ for two separable periodic systems with specified initial conditions $(J_1, \bar{\beta}_1, \bar{\beta}_2)$. The trajectory is given by Eqs. (12.35). The first orbit is periodic with period 2π in the horizontal direction. The second orbit, obtained through Eq. (12.36), is also 2π-periodic.

For the present problem, one can make an even simpler phase-space plot. Assume that the physical parameters of each system are given, including the angle (oscillator) frequencies (ω_1, ω_2). Assume also that the initial conditions $(J_1, \bar{\beta}_1, \bar{\beta}_2)$ are specified. Then the dynamical evolution of this collection of two separable, periodic systems is completely specified by $\sqrt{J_2/J_1}$ [Eqs. (12.32)]

110

and $\tilde{\phi}_1(t)$ [Eqs. (12.35)]; evidently

$$\tilde{\phi}_2(t) = \left(\frac{\omega_2}{\omega_1}\right)\tilde{\phi}_1(t) \tag{12.36}$$

In a two-dimensional coordinate system with $\sqrt{J_2/J_1}$ as ordinate and $\tilde{\phi}_1$ as abscissa, the orbit is simply a horizontal straight line! This situation is illustrated in Fig. 12.4.[70]

13   Perturbation of periodic hamiltonian systems

In the previous section we examined the behavior of separable, periodic hamiltonian systems and saw how action-angle variables provide the most direct and concise description of the dynamics. We now turn to the problem of the *coupling* among such systems. We shall focus first on the effect of adding a perturbation to one of them so that its hamiltonian becomes

$$H(p,q,t) \quad = \quad H_0(p,q) + \varepsilon V(p,q,t) \tag{13.1}$$

where ε is a small control parameter. In order to examine the effects of this perturbation, it is necessary to review some of the essential features of Hamilton-Jacobi theory for a single system as presented in Sec. 35.

Hamilton-Jacobi theory revisited

Consider first the problem with $\varepsilon = 0$. A canonical transformation from the pair of coordinates (p,q) to the new pair (P,Q) is obtained from the generating functions $S_0(q,P,t)$, where we now use a subscript zero for this unperturbed part of the problem. The corresponding momentum p and coordinate Q are obtained from this generating function through Eqs. (35.4)

$$p \quad = \quad \frac{\partial S_0(q,P,t)}{\partial q}$$

$$Q \quad = \quad \frac{\partial S_0(q,P,t)}{\partial P} \tag{13.2}$$

The new hamiltonian is given by Eq. (35.6)

$$\tilde{H}_0(P,Q,t) \quad = \quad H_0(p,q,t) + \frac{\partial S_0(q,P,t)}{\partial t} \tag{13.3}$$

---

[70] If the number ω_2/ω_1 is rational, it can be written as the ratio of two integers I_2/I_1. Let $\omega_1 t = 2\pi I_1$. Then $\omega_2 t = (\omega_2/\omega_1)\omega_1 t = 2\pi I_2$. After this time, all the angle variables in Eqs. (12.35) have gone through an integer number of 2π periods, and all the coordinates have thus returned to their initial values. The motion then exactly repeats itself. If the ratio ω_2/ω_1 is irrational, this exact repetition will never happen.

Since the transformation is canonical, Hamilton's equations are preserved

$$\dot{Q} = \frac{\partial \tilde{H}_0(P,Q,t)}{\partial P}$$

$$\dot{P} = -\frac{\partial \tilde{H}_0(P,Q,t)}{\partial Q} \qquad (13.4)$$

If one makes the clever choice of S_0 so that

$$\tilde{H}_0(P,Q,t) = 0 \qquad (13.5)$$

then the new set of canonical variables (P,Q) will be constants of the motion. The condition that the generating function produces this result is just the Hamilton-Jacobi Eq. (35.12)

$$H_0\left(\frac{\partial S_0}{\partial q}, q, t\right) + \frac{\partial S_0}{\partial t} = 0 \qquad (13.6)$$

Now add a perturbation $\varepsilon V(P,Q,t)$ as in Eq. (13.1), but retain the canonical transformation generated by $S_0(q,P,t)$.[71] In this case Eq. (13.3) becomes

$$\tilde{H}(P,Q,t) = H_0\left(\frac{\partial S_0}{\partial q}, q, t\right) + \varepsilon V(P,Q,t) + \frac{\partial S_0}{\partial t} \qquad (13.7)$$

We define

$$\tilde{H}(P,Q,t) = \tilde{H}_0(P,Q,t) + \tilde{H}_1(P,Q,t) \qquad (13.8)$$

with $\tilde{H}_1(P,Q,t) = \varepsilon V(P,Q,t)$. Use of Eqs. (13.5) and (13.6) then reduces these relations to

$$\tilde{H}(P,Q,t) = \varepsilon V(P,Q,t)$$
$$= \tilde{H}_1(P,Q,t) \qquad (13.9)$$

Hamilton's equations, which are preserved under the canonical transformation, now read

$$\dot{Q} = \frac{\partial \tilde{H}_1(P,Q,t)}{\partial P}$$

$$\dot{P} = -\frac{\partial \tilde{H}_1(P,Q,t)}{\partial Q} \qquad (13.10)$$

The canonical coordinates (P,Q) are no longer constants of the motion; instead, they are driven by the perturbation $\tilde{H}_1(P,Q,t)$ in Eq. (13.9). Evidently, the role of the canonical transformation generated by $S_0(q,P,t)$ has been to remove $\tilde{H}_0(P,Q,t)$ from the dynamics.[72]

---

[71]One can, as in Sec. 8, apply Hamilton-Jacobi theory to the full one-body hamiltonian. We focus here on developing perturbation theory in the small parameter ε.

[72]This observation plays a key role in subsequent developments.

Since $S_0(q, P, t)$ depends only on the one-body hamiltonian $H_0(p, q, t)$, the analysis of it proceeds exactly as in Sec. 8. In particular, for a conservative periodic system with $H_0(p, q)$, the new canonical momentum P can be identified with the action J, and the new coordinate $\bar{Q}$ (identical to $\bar{\beta}$) with the angle ϕ according to Eqs. (8.20), (8.22), and (8.24)

$$
\begin{aligned}
P &= J \\
\bar{Q} &= \bar{\beta} = \phi - \omega(J)t \\
\omega(J) &= \frac{dE_0(J)}{dJ}
\end{aligned}
\tag{13.11}
$$

The canonical transformation is from (p, q) to $(J, \bar{\beta})$, and Eq. (13.10) shows that J and $\bar{\beta}$ become weakly time dependent through terms of order ε.

We now make the following observation. Write the perturbation as $\varepsilon V(J, \phi, t)$. Since t and J are held constant in the partial differentiation, one can employ the rule for the differentiation of an implicit function to give

$$
\begin{aligned}
\frac{\partial \tilde{H}_1(J, \phi, t)}{\partial \bar{\beta}} &= \frac{\partial \tilde{H}_1(J, \phi, t)}{\partial \phi} \frac{\partial \phi}{\partial \bar{\beta}} \\
&= \frac{\partial \tilde{H}_1(J, \phi, t)}{\partial \phi}
\end{aligned}
\tag{13.12}
$$

and thus the second of Eqs. (13.10) becomes

$$
\dot{J} = -\frac{\partial \tilde{H}_1(J, \phi, t)}{\partial \phi}
\tag{13.13}
$$

To obtain a dynamical equation for ϕ, use the assumed functional form $\tilde{H}_1(J, \phi, t) = \tilde{H}_1[J, \bar{\beta} + \omega(J)t, t]$, and the requirement that the partial derivative is to be carried out at fixed $\bar{Q} = \bar{\beta}$ and t. One finds from the first of Eqs. (13.10)

$$
\begin{aligned}
\frac{d\bar{\beta}}{dt} &= \frac{\partial \tilde{H}_1[J, \bar{\beta} + \omega(J)t, t]}{\partial J} \\
&= \frac{\partial \tilde{H}_1(J, \phi, t)}{\partial J} + \frac{\partial \tilde{H}_1(J, \phi, t)}{\partial \phi} \frac{d\omega(J)}{dJ} t
\end{aligned}
\tag{13.14}
$$

By definition, however, the total time derivative of $\bar{\beta} = \phi - \omega(J)t$ in Eqs. (13.11) is given by

$$
\frac{d\bar{\beta}}{dt} = \frac{d\phi}{dt} - \omega(J) - \frac{d\omega(J)}{dJ} \frac{dJ}{dt} t
\tag{13.15}
$$

These two expressions for $d\bar{\beta}/dt$ can now be equated. The secular terms (those explicitly proportional to t) cancel with the aid of Eq. (13.13), and thus one has for the time derivative of the angle variable

$$
\dot{\phi} = \omega(J) + \frac{\partial \tilde{H}_1(J, \phi, t)}{\partial J}
\tag{13.16}
$$

113

Now $E_0(J)$ has no ϕ dependence, and its derivative with respect to J is just ω

$$\frac{\partial E_0(J)}{\partial \phi} = 0 \qquad ; \qquad \frac{\partial E_0(J)}{\partial J} = \omega(J) \qquad (13.17)$$

Thus Hamilton's Eqs. (13.10), originally written in terms of $(J, \bar{\beta})$, can be rewritten in terms of (J, ϕ) as

$$\dot{J} = -\frac{\partial H(J, \phi, t)}{\partial \phi}$$

$$\dot{\phi} = \frac{\partial H(J, \phi, t)}{\partial J}$$

$$H(J, \phi, t) = E_0(J) + \varepsilon V(J, \phi, t) \qquad (13.18)$$

We draw the following important conclusion: Even though the coordinates (J, ϕ) are not the familiar canonical pair $(J, \bar{\beta})$ that arise in solving the Hamilton-Jacobi equation, they differ from them only by the time translation in Eq. (13.11). Since t is held constant in the partial differentiation, the pair (J, ϕ) *does preserve Hamilton's equations* with the hamiltonian in Eq. (13.18) [compare Eqs. (12.30)]. This latter observation strongly suggests that (J, ϕ) also provide a canonical pair of variables.

This conclusion is easily verified. Consider the transformation [compare Eqs. (34.1)] from the Hamilton-Jacobi variables $(J, \bar{\beta})$ to a new set $(\tilde{J}, \phi)$

$$\tilde{J} = J$$

$$\phi = \omega(J)t + \bar{\beta} \qquad (13.19)$$

so that $\tilde{J}$ is always equal to J. It is easy to evaluate the Poisson bracket $[\tilde{J}, \phi]_{\text{PB}}$ in terms of the old variables $(J, \bar{\beta})$, which readily yields $[\tilde{J}, \phi]_{\text{PB}} = -1$.[73] As noted in Eqs. (37.1) and (37.9), this Poisson bracket is the criterion for a canonical transformation, so that the variables $(\tilde{J}, \phi)$ [the same as (J, ϕ)] are indeed a canonical pair.

To confirm that this transformation is canonical, we can also exhibit the following generator of a canonical transformation from the old variables $(J, \bar{\beta})$ to new ones $(\tilde{J}, \phi)$ (and it again turns out that $\tilde{J} = J$):

$$\tilde{S}_0(\bar{\beta}, \tilde{J}, t) = \bar{\beta}\tilde{J} + E_0(\tilde{J})t \qquad (13.20)$$

Equations (13.2) give the relations[74]

$$J = \frac{\partial \tilde{S}_0(\bar{\beta}, \tilde{J}, t)}{\partial \bar{\beta}} = \tilde{J} \qquad (13.21)$$

---

[73]The Poisson bracket as defined in Eq. (37.1) is given by

$$[\tilde{J}, \phi]_{\text{PB}} = \frac{\partial \tilde{J}(J, \bar{\beta}, t)}{\partial \bar{\beta}} \frac{\partial \phi(J, \bar{\beta}, t)}{\partial J} - \frac{\partial \tilde{J}(J, \bar{\beta}, t)}{\partial J} \frac{\partial \phi(J, \bar{\beta}, t)}{\partial \bar{\beta}} = -1$$

[74]The correspondence is $(p, q) \to (J, \bar{\beta})$ and $(P, Q) \to (\tilde{J}, \phi)$.

so that J is unchanged, and

$$\phi = \frac{\partial \tilde{S}_0(\bar{\beta}, \tilde{J}, t)}{\partial \tilde{J}} = \bar{\beta} + \frac{\partial E_0(\tilde{J})}{\partial \tilde{J}} t = \bar{\beta} + \omega(\tilde{J})t \qquad (13.22)$$

since $\omega(J) = dE_0(J)/dJ$. Equation (13.22) is just the usual relation $\phi = \bar{\beta} + \omega(J)t$. The corresponding new hamiltonian is [here, we follow the notation of Eqs. (13.18)]

$$H(J, \phi, t) = \tilde{H}_1(J, \phi, t) + \frac{\partial \tilde{S}_0(\bar{\beta}, \tilde{J}, t)}{\partial t} = E_0(J) + \varepsilon V(J, \phi, t) \qquad (13.23)$$

which is just the last of Eqs. (13.18). By construction, the new hamiltonian equations are the first two of Eqs. (13.18).

In direct analogy with the discussion in Sec. 32, Hamilton's Eqs. (13.18) can be derived from a modified Hamilton's principle

$$\delta \int_{t_1}^{t_2} L\, dt = \delta \int_{t_1}^{t_2} [J\dot{\phi} - H(J, \phi, t)]\, dt = 0 \qquad (13.24)$$

As in Sec. 34, one can then add a total time derivative to the integrand that generates a transformation to a new hamiltonian and new set of action-angle variables while preserving this form of Hamilton's equations. The preceding discussion shows explicitly that this transformation is canonical.

Equations (13.18) will form the starting point for our subsequent analysis of the effects of the coupling of separable, periodic hamiltonian systems. We give an example.

Direct perturbation analysis of the anharmonic oscillator

Consider the anharmonic oscillator of Sec. 7 with hamiltonian

$$H(p, q) = \frac{p^2}{2m} + \frac{1}{2}m\omega_0^2 q^2 + \varepsilon \frac{1}{4}mq^4 \qquad (13.25)$$

The relation between the coordinates (p, q) and the pair (J, ϕ), as generated by $\bar{S}_0(q, J, t)$ for the simple harmonic oscillator, is given by Eqs. (8.40)

$$p = \sqrt{2m\omega_0 J}\, \cos\phi$$
$$q = \sqrt{\frac{2J}{m\omega_0}}\, \sin\phi$$
$$\phi = \omega_0 t + \bar{\beta} \qquad (13.26)$$

The expression for $E_0(J)$ is given by Eq. (8.34)

$$E_0(J) = \omega_0 J \qquad (13.27)$$

The hamiltonian in Eq. (13.25) can thus be written in terms of (J, ϕ) as[75]

$$
\begin{aligned}
H(J, \phi) &= \omega_0 J + \varepsilon V(J, \phi) \\
V(J, \phi) &= \frac{m}{4}\left(\frac{2J}{m\omega_0}\right)^2 \sin^4 \phi = \frac{J^2}{m\omega_0^2}\sin^4 \phi
\end{aligned}
\qquad (13.28)
$$

Note the following important features of $V(J, \phi)$:

- It is explicitly *periodic* in the angle variable ϕ with period 2π;

- It is thus (necessarily) *nonlinear* in its dependence on ϕ;

- In this problem, it contains no further explicit t dependence.

Hamilton's Eqs. (13.18) now follow from this hamiltonian as

$$
\begin{aligned}
\dot{J} &= -\varepsilon \frac{4J^2}{m\omega_0^2}\sin^3 \phi \cos \phi \\
\dot{\phi} &= \omega_0 + \varepsilon \frac{2J}{m\omega_0^2}\sin^4 \phi
\end{aligned}
\qquad (13.29)
$$

The result is a set of coupled, nonlinear equations in the time t for the action-angle pair (J, ϕ), with a coupling strength ε. If $\varepsilon = 0$, the solution to these equations is just that in Eqs. (13.11) with constant $(J, \bar{\beta})$ and $\omega = \omega_0$. The corresponding (J, ϕ) phase-space orbit is just a horizontal straight line (compare Fig. 13.1). If $\varepsilon \neq 0$, the couplings in Eqs. (13.29) introduce oscillations of amplitude $O(\varepsilon)$ about this line.

We can be more quantitative by considering the average of these two dynamical Eqs. (13.29) over one cycle of the angle variable ϕ. Specifically, let $\langle \cdots \rangle = (2\pi)^{-1} \int_0^{2\pi} d\phi \cdots$. It is easy to see that $\langle \sin^3 \phi \cos \phi \rangle = 0$, so that $\langle \dot{J} \rangle$ vanishes. Thus the perturbed orbit includes small oscillations of J around its unperturbed value, but there is no net change over one cycle. In contrast, $\langle \sin^4 \phi \rangle = 3/8$, so that the net change in $\dot{\phi}$ now becomes

$$
\begin{aligned}
\langle \dot{\phi} \rangle &= \omega_0 + \frac{3\varepsilon J}{4m\omega_0^2} \\
&= \omega_0 + \frac{3\varepsilon a^2}{8\omega_0}
\end{aligned}
\qquad (13.30)
$$

where we use relation $J \approx E_0/\omega_0$ and the initial condition $E_0 = \frac{1}{2}m\omega_0^2 a^2$. By definition, the net increase in the angle variable ϕ over one cycle is 2π. In the present situation, it is also given approximately by $\langle \dot{\phi} \rangle \tau$, which differs from $\omega_0 \tau$

---

[75]Note that $\sin^4 \phi$ in the perturbation $V(J, \phi)$ can be expressed as a Fourier series in $\exp im\phi$ with $m = 0, \pm 2, \pm 4$, corresponding to the structure seen in Eq. (13.42) below.

by a correction of order ε. Comparison with Eq. (8.28) shows that the perturbed angular frequency is

$$\omega = \omega_0 + \frac{3\varepsilon a^2}{8\omega_0} \qquad (13.31)$$

to first order in ε. This expression is exactly the same as that found in Eq. (7.26) by eliminating the secular term from the equation of motion.

Model one-body problem with time-periodic perturbation

To investigate the coupling of separable, periodic systems we start with a model *one-body problem with a periodic, time-dependent interaction* [Ru86, Pe99]. In effect, we assume that another system with a different period has a close-lying orbit that interacts with the first one and affects its motion. Thus we add a one-body perturbation of the form $\varepsilon V(J, \phi, t)$ to the hamiltonian $E_0(J)$ of the first system and assume the following:[76]

- As in the previous example, it is assumed that $V(J, \phi, t)$ is periodic in the angle variable ϕ with period 2π;

- It is assumed that $V(J, \phi, t)$ is periodic in the time t with period τ_2. In fact, one can rescale the time dependence according to

$$t \rightarrow \frac{2\pi}{\tau_2} t \qquad (13.32)$$

so that it is no loss of generality to assume that V is 2π-periodic in t as well.

It is clear from Eqs. (13.18) that the perturbation $\varepsilon V(J, \phi, t)$ will put ripples on the straight-line, horizontal action-angle phase-space orbits obtained with $\varepsilon = 0$ (Fig. 13.1).

We have seen in Sec. 8 that it is possible to obtain the simple (J, ϕ) orbit in the unperturbed problem by carrying out a canonical transformation $\bar{S}_0(q, J, t)$ that eliminates $H_0(p, q)$ from the problem and leaves the time development of the coordinate in terms of the angle variable ϕ and the constant of the motion J (the action). In the present case of a perturbed one-body problem with Eq. (13.18), we can ask whether it is possible to find a *further canonical transformation to a new set of variables $(\bar{J}, \bar{\phi})$ that eliminates the wrinkles in the perturbed problem and restores the straight-line trajectory through $O(\varepsilon)$* (see Fig. 13.1). Can one straighten out the wrinkles? For most values of the parameters, we will find that it is indeed possible to carry out this procedure, but there are always special cases where one cannot do so.

---

[76]Note that with these assumptions, $V(J, \phi, t)$ will necessarily be nonlinear in both ϕ and t [see Eq. (13.42)].

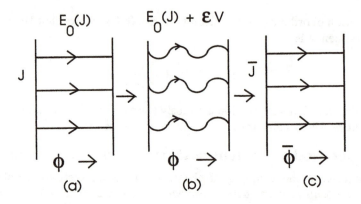

Fig. 13.1. (a) Unperturbed action-angle (J, ϕ) phase space. (b) Effect of adding a time-periodic perturbation of the form in Eq. (13.18) on originally unperturbed one-body action-angle (J, ϕ) phase space. (c) Result of additional canonical transformation to new variables $(\bar{J}, \bar{\phi})$ that will straighten the wrinkles to $O(\varepsilon^2)$.

We seek a transformation from (J, ϕ) to $(\bar{J}, \bar{\phi})$ where the new hamiltonian is simply

$$\bar{H}(\bar{J}, \bar{\phi}) = E_0(\bar{J}) + O(\varepsilon^2) \tag{13.33}$$

We know from Eq. (11.1) that the generator of the identity transformation is $\bar{J}\phi$. We include an additional contribution to the generator of $O(\varepsilon)$ that should produce this new hamiltonian. Thus we look for a new generating function

$$S_1(\bar{J}, \phi, t) = \bar{J}\phi + \varepsilon G(\bar{J}, \phi, t) \tag{13.34}$$

such that

$$\begin{aligned}
\bar{H}(\bar{J}, \bar{\phi}) &= E_0(J) + \varepsilon V(J, \phi, t) + \varepsilon \frac{\partial G(\bar{J}, \phi, t)}{\partial t} \\
&= E_0(\bar{J}) + O(\varepsilon^2)
\end{aligned} \tag{13.35}$$

As in Eq. (13.2), the new variables for this transformation are related to the old ones by[77]

$$\begin{aligned}
J &= \frac{\partial S_1(\bar{J}, \phi, t)}{\partial \phi} = \bar{J} + \varepsilon \frac{\partial G(\bar{J}, \phi, t)}{\partial \phi} \\
\bar{\phi} &= \frac{\partial S_1(\bar{J}, \phi, t)}{\partial \bar{J}} = \phi + \varepsilon \frac{\partial G(\bar{J}, \phi, t)}{\partial \bar{J}}
\end{aligned} \tag{13.36}$$

These equations can be rewritten as

$$\begin{aligned}
\bar{J} &= J - \varepsilon \frac{\partial G(\bar{J}, \phi, t)}{\partial \phi} \\
\bar{\phi} &= \phi + \varepsilon \frac{\partial G(\bar{J}, \phi, t)}{\partial \bar{J}}
\end{aligned} \tag{13.37}$$

[77]The correspondence in Eqs. (13.2) is $(p, q) \rightarrow (J, \phi)$ and $(P, Q) \rightarrow (\bar{J}, \bar{\phi})$; here, for notational clarity, we simply rearrange the order in which the variables are written in the generating function.

The transformed hamiltonian in Eq. (13.35) is expressed in terms of these transformed variables as

$$
\begin{aligned}
\bar{H}(\bar{J}, \bar{\phi}) &= E_0\left(\bar{J} + \varepsilon\frac{\partial G}{\partial \phi}\right) + \varepsilon V\left(\bar{J} + \varepsilon\frac{\partial G}{\partial \phi}, \phi, t\right) + \varepsilon\frac{\partial G}{\partial t} \\
&\equiv E_0(\bar{J}) + \varepsilon\frac{dE_0(\bar{J})}{d\bar{J}}\frac{\partial G}{\partial \phi} + \varepsilon V(\bar{J}, \phi, t) + \varepsilon\frac{\partial G}{\partial t} \\
&\quad + \varepsilon\left[V\left(\bar{J} + \varepsilon\frac{\partial G}{\partial \phi}, \phi, t\right) - V(\bar{J}, \phi, t)\right] \\
&\quad + \left[E_0\left(\bar{J} + \varepsilon\frac{\partial G}{\partial \phi}\right) - E_0(\bar{J}) - \varepsilon\frac{dE_0(\bar{J})}{d\bar{J}}\frac{\partial G}{\partial \phi}\right] \quad (13.38)
\end{aligned}
$$

Here the right hand side of the first equality has simply been rewritten as an identity by adding and subtracting the same terms. Now it is clear that each of the last two lines is explicitly of $O(\varepsilon^2)$. With the realization that

$$
\frac{dE_0(\bar{J})}{d\bar{J}} = \omega(\bar{J}) \tag{13.39}
$$

the condition that the terms of $O(\varepsilon)$ be eliminated in the second of Eqs. (13.38) reduces to[78]

$$
\omega(\bar{J})\frac{\partial G(\bar{J}, \phi, t)}{\partial \phi} + V(\bar{J}, \phi, t) + \frac{\partial G(\bar{J}, \phi, t)}{\partial t} = 0 \tag{13.40}
$$

Since Hamilton's Eqs. (13.18) are preserved under the canonical transformation, the new hamiltonian of Eq. (13.33) gives

$$
\begin{aligned}
\bar{H}(\bar{J}, \bar{\phi}) &= E_0(\bar{J}) + O(\varepsilon^2) \\
\frac{d\bar{J}}{dt} &= -\frac{\partial \bar{H}(\bar{J}, \bar{\phi})}{\partial \bar{\phi}} = 0 + O(\varepsilon^2) \\
\frac{d\bar{\phi}}{dt} &= \frac{\partial \bar{H}(\bar{J}, \bar{\phi})}{\partial \bar{J}} = \omega(\bar{J}) + O(\varepsilon^2) \tag{13.41}
\end{aligned}
$$

Thus the orbits in the $(\bar{J}, \bar{\phi})$ plane have indeed been straightened through $O(\varepsilon)$ (see Fig. 13.1). It remains to find the solution to Eq. (13.40).

Since $V(\bar{J}, \phi, t)$ is periodic with period 2π in both ϕ and the (rescaled) time t, one can, with no loss of generality, make the double Fourier series expansion

$$
V(\bar{J}, \phi, t) = {\sum_{m,n}}' V_{mn}(\bar{J})e^{im\phi}e^{-int} \tag{13.42}
$$

For most problems of interest, there will be only a few terms in this sum. The amplitudes $V_{mn}(\bar{J})$ are now known quantities. One can assume that $V_{00}(\bar{J}) = 0$, for any such constant term is readily included in $E_0(\bar{J})$. We shall keep the

---

[78]If this condition holds for all ϕ, it will also be true when ϕ is replaced by $\bar{\phi}$.

omission of this term in mind by putting a prime on the double sum. This form of V suggests that one look for a solution to Eq. (13.40) with the same basic structure

$$G(\bar{J}, \phi, t) = \sum_{m,n}' G_{mn}(\bar{J}) e^{im\phi} e^{-int} \tag{13.43}$$

Substitution of these expansions in Eq. (13.40) yields

$$\sum_{m,n}' \{ [im\, \omega(\bar{J}) - in] G_{mn}(\bar{J}) + V_{mn}(\bar{J}) \} e^{im\phi} e^{-int} = 0 \tag{13.44}$$

The linear independence of the exponentials implies that their coefficients must vanish. Thus

$$G_{mn}(\bar{J}) = \frac{i V_{mn}(\bar{J})}{m\, \omega(\bar{J}) - n}$$

$$G(\bar{J}, \phi, t) = \sum_{m,n}' \frac{i V_{mn}(\bar{J})}{m\, \omega(\bar{J}) - n} e^{im\phi} e^{-int} \tag{13.45}$$

Now if

$$m\omega(\bar{J}) - n = 0 \qquad ; \text{ any } (m, n) \text{ in sum} \tag{13.46}$$

there is a problem, for the denominator in Eq. (13.45) then vanishes. If this were never to happen, then Eq. (13.45) provides an explicit solution for the generator, and the action-angle orbits in $(\bar{J}, \bar{\phi})$ have been straightened to $O(\varepsilon^2)$, as indicated in Eqs. (13.41) and Fig. 13.1. Typically, however, $\omega(\bar{J})$ will depend on $\bar{J}$, and as $\bar{J}$ varies, one will *always* run into problems with the vanishing denominator.[79]

Resonant disruption of phase space

In order to understand the consequences of a vanishing denominator, we examine a simple model where the interaction has only a single Fourier component

$$H(J, \phi, t) = E_0(J) + \varepsilon f(J) \cos(m\phi - nt) \tag{13.47}$$

and assume that the action J is close to the resonant value J_r where[80]

$$m\omega(J_r) - n = 0 \tag{13.48}$$

Now perform the following:
1). Carry out a *new* canonical transformation on this hamiltonian with the generator[81]

$$\tilde{S}_1(\tilde{J}, \phi, t) = \tilde{J}\left(\phi - \frac{nt}{m}\right) \tag{13.49}$$

---

[79] It is only for the linear oscillator that $\omega = \omega_0$ is independent of $\bar{J}$.

[80] Note that the condition $\omega(J_r) = n/m$ implies that the angular frequency takes a certain *rational* value.

[81] This canonical transformation is unrelated to that with the generator G in Eq. (13.34). To avoid any possible confusion, we here use tildes to denote the transformed variables.

The transformed variables are

$$J = \frac{\tilde{S}_1(\tilde{J}, \phi, t)}{\partial \phi} = \tilde{J}$$

$$\tilde{\phi} = \frac{\tilde{S}_1(\tilde{J}, \phi, t)}{\partial \tilde{J}} = \phi - \frac{nt}{m} \tag{13.50}$$

This transformation leaves J unchanged and simply moves $\tilde{\phi}$ relative to ϕ by a linear shift in the time. The new hamiltonian is then

$$\tilde{H}(\tilde{J}, \tilde{\phi}, t) = H(J, \phi, t) + \frac{\tilde{S}_1(\tilde{J}, \phi, t)}{\partial t} \tag{13.51}$$

where the right hand side is re-expressed in terms of the new variables. Thus we have (here we can simply set $\tilde{J} = J$)

$$\tilde{H}(J, \tilde{\phi}) = E_0(J) + \varepsilon f(J) \cos m\tilde{\phi} - \frac{nJ}{m} \qquad ; \tilde{J} = J \tag{13.52}$$

2). Make a series expansion in J near the resonant value J_r

$$\begin{aligned}
E_0(J) &= E_0(J_r) + \left(\frac{dE_0}{dJ}\right)_{J_r} (J - J_r) + \frac{1}{2}\left(\frac{d^2 E_0}{dJ^2}\right)_{J_r} (J - J_r)^2 + \cdots \\
&= E_0(J_r) + \omega(J_r)(J - J_r) + \frac{1}{2}\omega'(J_r)(J - J_r)^2 + \cdots \\
f(J) &= f(J_r) + f'(J_r)(J - J_r) + \cdots
\end{aligned} \tag{13.53}$$

It is sufficient to retain only the first term in the last expression since its contribution in Eq. (13.52) is already of $O(\varepsilon)$. Insertion of these expansions in Eq. (13.52) and the use of Eq. (13.48) gives

$$\tilde{H}(J, \tilde{\phi}) = E_0(J_r) + \frac{1}{2}\omega'(J_r)(J - J_r)^2 + \varepsilon f(J_r) \cos m\tilde{\phi} - \frac{n}{m}J_r + \cdots \tag{13.54}$$

The canonical transformation in Eq. (13.49) has eliminated the linear term in $(J - J_r)$ in this expression to $O(\varepsilon)$.

The new hamiltonian in Eq. (13.54) has a simple structure

$$\tilde{H}(J, \tilde{\phi}) \doteq \frac{1}{2}\omega'(J_r)(J - J_r)^2 + \varepsilon f(J_r) \cos m\tilde{\phi} \tag{13.55}$$

where the symbol $\doteq$ implies that the (irrelevant) constant terms have been dropped. This is just the *pendulum hamiltonian*, and one can write it as

$$\begin{aligned}
\tilde{H}(J, \tilde{\phi}) &\doteq \frac{\mathcal{J}^2}{2I} + M_0 g \cos m\tilde{\phi} \\
\mathcal{J} &= J - J_r \\
M_0 g &\equiv \varepsilon f(J_r) \qquad ; \omega'(J_r) \equiv \frac{1}{I}
\end{aligned} \tag{13.56}$$

Here M_0 is an effective mass (we use M_0 to distinguish it from the integer m), and it is assumed here that the last two quantities are positive. The canonical transformation preserves Hamilton's equations in the new variables $(\tilde{J}, \tilde{\phi}) = (J, \tilde{\phi})$, which are equivalent to $(\mathcal{J}, \tilde{\phi})$ since only a constant shift in J is involved. One recovers the equations of motion for the simple pendulum.

$$\tilde{H}(J, \tilde{\phi}) \;\dot{=}\; \frac{\mathcal{J}^2}{2I} + M_0 g \cos m\tilde{\phi}$$

$$\frac{dJ}{dt} \;=\; M_0 g\, m \sin m\tilde{\phi}$$

$$\frac{d\tilde{\phi}}{dt} \;=\; \frac{1}{I}\mathcal{J} \tag{13.57}$$

and all the one-body analysis of Sec. 8 then applies.

With the signs as given, the pendulum is *up* at $m\tilde{\phi} = \pm 2\pi k$, and it is *down* at $m\tilde{\phi} = \pm\pi(2k+1)$ where $k = 0, 1, 2, \cdots$. It is evidently *unstable* at $\mathcal{J} = 0$ (or $J = J_r$) in the up case and *stable* in the down case. Thus at $J = J_r$

$$\tilde{\phi} \;=\; \pm\frac{\pi}{m}, \pm\frac{3\pi}{m}, \cdots \qquad ; \text{down and stable}$$

$$\;=\; 0, \pm\frac{2\pi}{m}, \pm\frac{4\pi}{m}, \cdots \qquad ; \text{up and unstable} \tag{13.58}$$

This leads to the simple and elegant $(J, \tilde{\phi})$ phase-space plot for $J \approx J_r$ shown in Fig. 13.2 for the particular case $m = 3$.

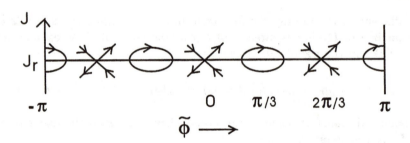

Fig. 13.2. Action-angle $(J, \tilde{\phi})$ phase-space plot, with hamiltonian of Eq. (13.47), for $J \approx J_r$ and $m = 3$; based on Eqs. (13.57) and (13.58). Here $\tilde{\phi} = \phi - nt/m$.

We are now in a position to sketch the full set of $(J, \tilde{\phi})$ phase-space orbits for the model problem in Eq. (13.47), and the result is shown in Fig. 13.3. Here J is not conserved, so the curves cannot be labeled with J. Away from $J \approx J_r$, one gets simple oscillatory wiggles that can be smoothed out through $O(\varepsilon)$ with the canonical transformation in Eq. (13.45), as previously discussed. Near J_r, however, the perturbation in Eq. (13.47) causes an *essential disruption* of phase space, and one has stable and unstable fixed points with islands of stability immersed in the plots.

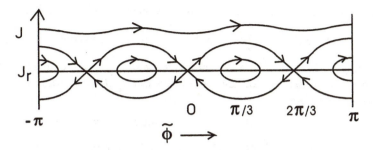

Fig. 13.3. Sketch of the action-angle $(J, \tilde{\phi})$ phase-space orbits with the hamiltonian of Eq. (13.47) for J near the resonant value J_r and $m = 3$.

The physical interpretation is the following. Consider a separable periodic system. The addition of the model nonlinear one-body time-periodic perturbation in Eq. (13.47), caused, for example, by a second interacting orbit, can lead to a resonance where

$$\omega(J_r) = \left[\frac{dE_0(J)}{dJ}\right]_{J_r} = \frac{n}{m} \tag{13.59}$$

Here the integer m characterizes the periodic ϕ dependence of the perturbation, and n the periodic t dependence of the perturbation. At the resonant action J_r, the stable fixed points of Eq. (13.58) will occur in the angle variable of Eq. (13.50)

$$\tilde{\phi} = \phi - \frac{n}{m}t = \phi - \omega(J_r)t$$

$$\tilde{\phi}_{\text{stable}} = \pm\frac{\pi}{m}, \pm\frac{3\pi}{m}, \cdots \tag{13.60}$$

In the vicinity of these stable fixed points, which occur on each side of $\tilde{\phi} = 0$ [or $\phi = \omega(J_r)t$], there are trapped orbits in the effective pendulum potential where the action and angle variables $(J, \tilde{\phi})$ perform small oscillations around the stable values.[82] There will similarly be a periodic array of unstable fixed points. This structure yields the disrupted region of phase space. For values of the action away from J_r, the originally uniform orbital motion in Eq. (13.11) is perturbed in a smooth manner, and phase space is undisrupted. The two phase-space regions are divided by the two separatrix orbits passing through the unstable fixed points.

A basic question concerns the *size* of the disrupted regions in phase space. This size can be estimated by studying the orbit that goes through the unstable

---

[82]The strength of the trapping potential is $\propto \varepsilon$ [Eq. (13.56)], but in classical mechanics, "trapped is trapped." In a statistical ensemble of particles over the perturbed single-particle orbits, there will be a one-dimensional periodic array of trapped islands of particles at the resonant frequency $\omega(J_r)$ and angular locations $\tilde{\phi}_{\text{stable}}$. They look like beads on a necklace.

fixed points, for the disrupted orbits and islands of stability are all are enclosed by the separatrix. Go back to the hamiltonian near J_r in Eq. (13.54)

$$\tilde{H}(\tilde{\phi}, J) \doteq \frac{1}{2}\omega'(J_r)(J - J_r)^2 + \varepsilon f(J_r)\cos m\tilde{\phi} \qquad (13.61)$$

Equation (13.61) describes a conservative system, so one can label the orbits by the value of $\tilde{H}$. At an unstable fixed point, $J = J_r$ and $\cos m\tilde{\phi} = 1$ [Eq. (13.58)]. Therefore[83]

$$\tilde{H} = \varepsilon f(J_r) \qquad ; \text{ unstable fixed point}$$

$$\varepsilon f(J_r) = \frac{1}{2}\omega'(J_r)(J - J_r)^2 + \varepsilon f(J_r)\cos m\tilde{\phi} \qquad (13.62)$$

This relation can be solved to find the separatrix curves $J(\tilde{\phi})$

$$J = J_r \pm \left[\frac{2\varepsilon f(J_r)(1 - \cos m\tilde{\phi})}{\omega'(J_r)}\right]^{1/2} \qquad (13.63)$$

where $\pm$ describe the upper and lower separatrix, respectively. The "size" of the stable island is evidently $\propto \sqrt{\varepsilon}$

$$|J - J_r| \propto \sqrt{\varepsilon} \qquad ; \text{ size of stable island} \qquad (13.64)$$

With a typical nonlinear system, $\omega(J)$ varies with J. If there are terms with additional values of (m, n) contained in the sum in Eq. (13.42), then additional resonances at $\omega(J_r) = n/m$ will occur. Thus phase space breaks into two different types of regions — resonant and nonresonant. The disrupted, resonant regions correspond to the (rational) values of $\omega(J_r) = n/m$ for (m, n) in the sum, while the smooth, nonresonant regions fill the rest of phase space. The situation is sketched schematically in Fig. (13.4). In most of phase space, one gets smooth regular motion.

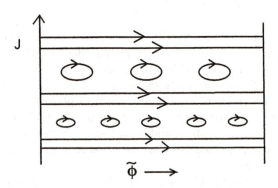

Fig. 13.4. Schematic sketch of one-body action-angle phase space with time-periodic perturbation of Eq. (13.42), showing smooth, nonresonant and disrupted, resonant regions.

---

[83]We restore the equality.

Overlap of disrupted regions

To estimate the extent of the phase-space disruption, it is necessary to have some criterion to ensure that the disrupted regions do not overlap. Assume that we are dealing with a finite set of Fourier components (m, n) in the perturbation in Eq. (13.42). Start from $\omega(J_r) = n/m$. There will then be some finite spacing $\Delta(n/m)$ for these ratios. The next resonance at $J_r + \Delta J_r$ will occur at $(n/m)' = n/m + \Delta(n/m)$. Thus

$$
\begin{aligned}
\omega(J_r + \Delta J_r) &= \frac{n}{m} + \omega'(J_r)\, \Delta J_r + \cdots \\
&= \frac{n}{m} + \Delta\left(\frac{n}{m}\right) + \cdots
\end{aligned}
\tag{13.65}
$$

Hence one has for the *spacing* between adjacent resonances

$$
\Delta J_r \approx \frac{\Delta(n/m)}{\omega'(J_r)}
\tag{13.66}
$$

An expression for the *width* of the disrupted regions can be obtained from Eq. (13.63)[84]

$$
\delta J_r \approx 2\left[\frac{2\varepsilon f(J_r)}{\omega'(J_r)}\right]^{1/2}
\tag{13.67}
$$

The condition that the disrupted regions not overlap is that their width be less than their spacing

$$
\begin{aligned}
\delta J_r &< \Delta J_r \\
[\varepsilon \omega'(J_r) f(J_r)]^{1/2} &< \frac{1}{2\sqrt{2}} \Delta\left(\frac{n}{m}\right)
\end{aligned}
\tag{13.68}
$$

This inequality is known as the *Chirikov criterion*: it ensures that the disrupted regions do not overlap [Ch79, Ru86, Pe99]. For a finite set of Fourier components (m, n) in the perturbation in Eq. (13.42), this criterion is always satisfied as $\varepsilon \to 0$.

Of course, all these arguments are based on the nonlinear one-body problem with the time-periodic perturbation in Eq. (13.42). The extension of these ideas to the actual coupling of separable period systems, and to higher dimensions, is the topic of the final section.

14 Coupled separable periodic hamiltonian systems

In Sec. 13, we discussed the dynamics of a a single nonlinear periodic system with a perturbation $\varepsilon V(J, \phi, t)$ that is 2π-periodic in both (ϕ, t). This one-body perturbation serves to model the effect of a second nearby system with

---

[84]Note it is assumed that $V_{mn}(J) \approx f(J)$ and $\omega'(J) > 0$ in this argument.

a different period. We now turn to the more realistic problem of two weakly coupled separable, periodic systems [Ru86, Pe99].[85]

Two weakly coupled degrees of freedom

With the discussion in Sec. 12 as background, we assume an uncoupled, zero-order action-angle hamiltonian of the form

$$H_0(J_1, J_2) = \sum_{\sigma=1}^{2} E_0(J_\sigma)_\sigma \qquad (14.1)$$

Here, the two actions J_1 and J_2 are constants of the motion, and H_0 and the dynamics are separately periodic in each angle variable (ϕ_1, ϕ_2) with period 2π. The phase trajectory therefore lies in the four-dimensional space $(J_1, J_2, \phi_1, \phi_2)$. The entire system is conservative, and one can work at fixed total energy E. This constraint confines the motion to a three-dimensional subspace. Within this subspace of given E, we can choose as coordinates the angles (ϕ_1, ϕ_2) and some nontrivial combination of (J_1, J_2), say J_2/J_1.

The unperturbed Hamilton's equations for this two-body system are[86]

$$\dot{J}_\sigma = -\frac{\partial H_0(J_1, J_2)}{\partial \phi_\sigma} = 0 \qquad\qquad ; \sigma = 1, 2$$

$$\dot{\phi}_\sigma = \frac{\partial H_0(J_1, J_2)}{\partial J_\sigma} = \frac{\partial E_0(J_\sigma)_\sigma}{\partial J_\sigma} = \omega_\sigma(J_\sigma) \qquad (14.2)$$

These equations yield a phase trajectory that is a simple flow in any given plane of fixed J_2/J_1, as illustrated in Fig. 14.1. In each plane the motion advances at an angle determined by the particular ratio ω_2/ω_1.

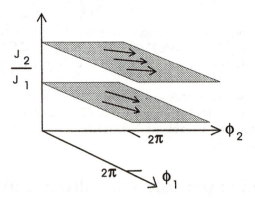

Fig. 14.1. Sketch of the phase-space trajectories for the conservative system of two uncoupled, separable periodic systems in the subspace of fixed total $E = E_0$ with action-angle coordinates $(\phi_1, \phi_2, J_2/J_1)$.

---

[85]In general, a time-independent separable hamiltonian system with N degrees of freedom can be reduced to a time-dependent periodic one with $N-1$ degrees of freedom [Jo98a].

[86]In this case, $E = E_0 = \sum_{\sigma=1}^{2} E_0(J_\sigma)_\sigma$.

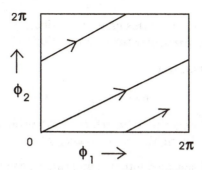

Fig. 14.2. An orbit in one of the planes in Fig. 14.1. It is periodic with period 2π in both the vertical and horizontal directions.

Each plane, which we further represent in Fig. 14.2, is topologically equivalent to a two-torus. Thus the situation can also be viewed as a collection of nested tori, with coordinates defined as in Fig. 14.3. On any given torus, we have

$$
\begin{aligned}
\dot{J_1} &= \dot{J_2} = 0 \\
\dot{\phi_1} &= \omega_1(J_1) \\
\dot{\phi_2} &= \omega_2(J_2)
\end{aligned}
\tag{14.3}
$$

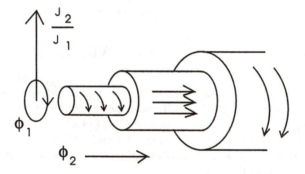

Fig. 14.3. Schematic redrawing of Fig. 14.1 as a series of nested tori (compare Fig. 12.2).

We note the following features of this result:

- Within the subspace of given total energy E, the initial conditions determine how E is divided between the two uncoupled dynamical systems $E = E_{01} + E_{02}$. Correspondingly, this apportionment fixes the actions $J_1(E_{01})$ and $J_2(E_{02})$, which in turn give both the ratio of the actions J_2/J_1 and the ratio of frequencies $\omega_2(J_2)/\omega_1(J_1)$;

- If the ratio ω_2/ω_1 is irrational (namely, the frequencies are incommensurate), the motion is called "quasiperiodic" because there are two *different*

127

fundamental frequencies. In this case, the trajectory eventually covers the entire surface of the particular torus in question;

- If the ratio is rational, so that

$$m\omega_1 - n\omega_2 = 0 \qquad (14.4)$$

for some specified integer pair (m, n), then the orbit will eventually close on itself and the entire motion is periodic;

- For any finite set of specified integer pairs (m, n), as relevant in the following discussion, the relation in Eq. (14.4) will be satisfied only on special tori.

Now suppose one turns on a weak *perturbative interaction* between the systems, of the form

$$H_1 = \varepsilon V(J_1, J_2, \phi_1, \phi_2) \qquad (14.5)$$

As in our previous examples, V is assumed to be 2π-periodic in both angle variables (ϕ_1, ϕ_2). The two-body extension of Hamilton's Eqs. (13.18) is now

$$\dot{J}_\sigma = -\frac{\partial H(J_1, J_2, \phi_1, \phi_2)}{\partial \phi_\sigma}$$

$$\dot{\phi}_\sigma = \frac{\partial H(J_1, J_2, \phi_1, \phi_2)}{\partial J_\sigma} \qquad ; \sigma = 1, 2$$

$$H(J_1, J_2, \phi_1, \phi_2) = \sum_{\sigma=1}^{2} E_0(J_\sigma)_\sigma + \varepsilon V(J_1, J_2, \phi_1, \phi_2) \qquad (14.6)$$

In contrast to the one-body model considered in Sec. 13, the coupling term $\varepsilon V(J_1, J_2, \phi_1, \phi_2)$ in the present hamiltonian has no explicit time dependence. If $\varepsilon = 0$, the equations of motion reduce to Eqs. (14.3). If $\varepsilon \neq 0$, the coupling term εV puts ripples on the orbits in Figs. 14.1 and 14.3.[87] The question arises, can one again find a canonical transformation that smooths out the wrinkles through $O(\varepsilon)$ (compare Fig. 13.1)?

Similar to Sec. 13, we attempt to accomplish this goal with a canonical transformation of the form[88]

$$S_1(\bar{J}_1, \bar{J}_2, \phi_1, \phi_2) = \sum_{\sigma=1}^{2} \bar{J}_\sigma \phi_\sigma + \varepsilon G(\bar{J}_1, \bar{J}_2, \phi_1, \phi_2) \qquad (14.7)$$

The two-body extension of the condition in Eq. (13.40) that the terms of $O(\varepsilon)$

---

[87] Since (J_1, J_2) now vary, the ripples take the trajectories out of the surfaces of fixed J_2/J_1.

[88] In Sec. 13, the time dependence of the transformation function in Eq. (13.34) served to mimic the coupling to another periodic system. Here, in contrast, the coupling involves simply the two angle variables (ϕ_1, ϕ_2), and the generating function S_1 is assumed to be independent of t.

disappear from the transformed hamiltonian is

$$\sum_{\sigma=1}^{2} \omega_\sigma(\bar{J}_\sigma) \frac{\partial G(\bar{J}_1, \bar{J}_2, \phi_1, \phi_2)}{\partial \phi_\sigma} + V(\bar{J}_1, \bar{J}_2, \phi_1, \phi_2) = 0 \qquad (14.8)$$

Since V is assumed to be periodic with period 2π in both angle variables, it has a double Fourier series expansion[89]

$$V(\bar{J}_1, \bar{J}_2, \phi_1, \phi_2) = {\sum_{m,n}}' V_{mn}(\bar{J}_1, \bar{J}_2) e^{im\phi_1} e^{-in\phi_2} \qquad (14.9)$$

This periodicity suggests that G has the same form, with no explicit dependence on the time t

$$G(\bar{J}_1, \bar{J}_2, \phi_1, \phi_2) = {\sum_{m,n}}' G_{mn}(\bar{J}_1, \bar{J}_2) e^{im\phi_1} e^{-in\phi_2} \qquad (14.10)$$

As in Eqs. (13.45), substitution in Eq. (14.8) leads to the solution

$$G_{mn}(\bar{J}_1, \bar{J}_2) = \frac{iV_{mn}(\bar{J}_1, \bar{J}_2)}{m\omega_1(\bar{J}_1) - n\omega_2(\bar{J}_2)}$$

$$G(\bar{J}_1, \bar{J}_2, \phi_1, \phi_2) = {\sum_{m,n}}' \frac{iV_{mn}(\bar{J}_1, \bar{J}_2)}{m\omega_1(\bar{J}_1) - n\omega_2(\bar{J}_2)} e^{im\phi_1} e^{-in\phi_2} \qquad (14.11)$$

This result provides an explicit solution that smooths out the wrinkles through $O(\varepsilon)$, *except when (i) the denominator vanishes, and (ii) the corresponding factor* $V_{mn}(\bar{J}_1, \bar{J}_2)$ *is nonzero.* The condition that the denominator vanish

$$m\omega_1(\bar{J}_{1r}) - n\omega_2(\bar{J}_{2r}) = 0 \qquad (14.12)$$

is precisely Eq. (14.4). At that frequency ratio

$$\frac{\omega_1(\bar{J}_{1r})}{\omega_2(\bar{J}_{2r})} = \frac{n}{m} \qquad (14.13)$$

there will again be a resonant disruption of phase space.

Resonant disruption of phase space

As in the previous section, we keep just one term in Eq. (14.9) and ask what happens in the vicinity of the resonance. Consider the model

$$H(J_1, J_2, \phi_1, \phi_2) = \sum_{\sigma=1}^{2} E_0(J_\sigma)_\sigma + \varepsilon f(J_1, J_2) \cos(m\phi_1 - n\phi_2) \qquad (14.14)$$

---

[89]The prime in the sum again indicates that any $(0,0)$ term can be incorporated into H_0; we assume that such a term is absent.

for $(J_1, J_2) \approx (J_{1r}, J_{2r})$. At resonance, we have the condition

$$\frac{\omega_1(J_{1r})}{\omega_2(J_{2r})} = \frac{n}{m} \qquad (14.15)$$

Introduce a *new* canonical transformation generated by [compare Eq. (13.49)]

$$\tilde{S}_1(\tilde{J}_1, \tilde{J}_2, \phi_1, \phi_2) = \left(\phi_1 - \frac{n}{m}\phi_2\right)\tilde{J}_1 + \phi_2\tilde{J}_2 \qquad (14.16)$$

The transformed variables are

$$J_1 = \frac{\partial\tilde{S}_1}{\partial\phi_1} = \tilde{J}_1 \qquad ; J_2 = \frac{\partial\tilde{S}_1}{\partial\phi_2} = \tilde{J}_2 - \frac{n}{m}\tilde{J}_1$$

$$\tilde{\phi}_1 = \frac{\partial\tilde{S}_1}{\partial\tilde{J}_1} = \phi_1 - \frac{n}{m}\phi_2 \qquad ; \tilde{\phi}_2 = \frac{\partial\tilde{S}_1}{\partial\tilde{J}_2} = \phi_2 \qquad (14.17)$$

and the new hamiltonian is

$$\tilde{H}(\tilde{J}_1, \tilde{J}_2, \tilde{\phi}_1, \tilde{\phi}_2) = H(J_1, J_2, \phi_1, \phi_2)$$

$$= \sum_{\sigma=1}^{2} E_0(J_\sigma)_\sigma + \varepsilon f(J_1, J_2)\cos(m\phi_1 - n\phi_2) \qquad (14.18)$$

with the right hand side expressed in the new variables.

Now expand each one-body term in the unperturbed hamiltonian around the resonant action, as in Eq. (13.53)

$$E_0(J) = E_0(J_r) + \omega(J_r)(J - J_r) + \frac{1}{2}\omega'(J_r)(J - J_r)^2 + \cdots \quad (14.19)$$

which makes the J dependence explicit. Substitution of this expansion and Eqs. (14.17) into Eq. (14.18) then gives the new hamiltonian[90]

$$\tilde{H}(\tilde{J}_1, \tilde{J}_2, \tilde{\phi}_1, \tilde{\phi}_2) = E_0(J_{1r})_1 + \omega_1(J_{1r})(\tilde{J}_1 - J_{1r}) + \frac{1}{2}\omega_1'(J_{1r})(\tilde{J}_1 - J_{1r})^2 +$$

$$E_0(J_{2r})_2 + \omega_2(J_{2r})\left(\tilde{J}_2 - \frac{n}{m}\tilde{J}_1 - J_{2r}\right) + \frac{1}{2}\omega_2'(J_{2r})\left(\tilde{J}_2 - \frac{n}{m}\tilde{J}_1 - J_{2r}\right)^2 +$$

$$\varepsilon f(J_{1r}, J_{2r})\cos m\tilde{\phi}_1 \qquad (14.20)$$

Use of the resonance condition in Eq. (14.15) eliminates the linear term in $\tilde{J}_1$ from this expression, so that

$$\tilde{H}(\tilde{J}_1, \tilde{J}_2, \tilde{\phi}_1, \tilde{\phi}_2) = E_0(J_{1r})_1 + \omega_1(J_{1r})(-J_{1r}) + \frac{1}{2}\omega_1'(J_{1r})(\tilde{J}_1 - J_{1r})^2 +$$

$$E_0(J_{2r})_2 + \omega_2(J_{2r})\left(\tilde{J}_2 - J_{2r}\right) + \frac{1}{2}\omega_2'(J_{2r})\left(\tilde{J}_2 - \frac{n}{m}\tilde{J}_1 - J_{2r}\right)^2 +$$

$$\varepsilon f(J_{1r}, J_{2r})\cos m\tilde{\phi}_1 \qquad (14.21)$$

[90]The quantity $f(J_1, J_2)$ can again be evaluated at resonance because it enters multiplied by ε.

The canonical transformation in Eq. (14.16) preserves the form of Hamilton's equations. In terms of the new variables, they now read

$$\frac{d\tilde{J}_1}{dt} = -\frac{\partial \tilde{H}(\tilde{J}_1, \tilde{J}_2, \tilde{\phi}_1, \tilde{\phi}_2)}{\partial \tilde{\phi}_1} = \varepsilon f(J_{1r}, J_{2r}) m \sin m\tilde{\phi}_1$$

$$\frac{d\tilde{J}_2}{dt} = -\frac{\partial \tilde{H}}{\partial \tilde{\phi}_2} = 0$$

$$\frac{d\tilde{\phi}_1}{dt} = \frac{\partial \tilde{H}}{\partial \tilde{J}_1} = \omega_1'(J_{1r})(\tilde{J}_1 - J_{1r}) - \frac{n}{m}\omega_2'(J_{2r})\left(\tilde{J}_2 - \frac{n}{m}\tilde{J}_1 - J_{2r}\right)$$

$$\frac{d\tilde{\phi}_2}{dt} = \frac{\partial \tilde{H}}{\partial \tilde{J}_2} = \omega_2(J_{2r}) + \omega_2'(J_{2r})\left(\tilde{J}_2 - \frac{n}{m}\tilde{J}_1 - J_{2r}\right) \qquad (14.22)$$

The second equation implies that $\tilde{J}_2$ is a constant of the motion. From Eqs. (14.17) $\tilde{J}_2$ is expressed in terms of (J_1, J_2) by

$$\tilde{J}_2 = J_2 + \frac{n}{m}J_1 \qquad (14.23)$$

Evaluation of this expression at resonance yields

$$\tilde{J}_2 = J_{2r} + \frac{n}{m}J_{1r} \qquad (14.24)$$

With this fixed value for $\tilde{J}_2$, Hamilton's equations take the simple form [note that $\tilde{J}_1 = J_1$ and $\tilde{\phi}_2 = \phi_2$ from Eqs. (14.17)]

$$\frac{d\,\delta J_1}{dt} = \varepsilon f(J_{1r}, J_{2r}) m \sin m\tilde{\phi}_1$$

$$\frac{d\tilde{\phi}_1}{dt} = a\,\delta J_1$$

$$\frac{d\phi_2}{dt} = \omega_2(J_{2r}) + b\,\delta J_1 \qquad ;\ \delta J_1 \equiv J_1 - J_{1r} \qquad (14.25)$$

where (a, b) are constants given by

$$a = \omega_1'(J_{1r}) + \left(\frac{n}{m}\right)^2 \omega_2'(J_{2r})$$

$$b = -\frac{n}{m}\omega_2'(J_{2r}) \qquad (14.26)$$

The first two of Eqs. (14.25) are just the *pendulum equations* for $(\tilde{\phi}_1, \delta J_1)$ [compare Eqs. (13.56) and (13.57)]. They may be solved together to produce the familiar phase-space orbits (compare Figs. 13.2 and 13.3). To zero order in ε, the angle ϕ_2 grows at a constant mean rate $\omega_2(J_{2r})\,t$, but this simple behavior is modified by the small variations of $O(\varepsilon)$ arising from the previously determined $\delta J_1(t)$. Here

$$\tilde{\phi}_1 = \phi_1 - \frac{n}{m}\phi_2$$

$$J_2 = \tilde{J}_2 - \frac{n}{m}J_1 = J_{2r} - \frac{n}{m}\delta J_1 \qquad (14.27)$$

131

so that the action J_2 also experiences small perturbations of order ε from $\delta J_1(t)$. Note that the first-order change in the ratio J_2/J_1 from the resonant ratio is now

$$\delta\left(\frac{J_2}{J_1}\right) = -\frac{1}{J_{1r}}\left(\frac{n}{m} + \frac{J_{2r}}{J_{1r}}\right)\delta J_1 \tag{14.28}$$

Near the resonant ratio $(J_2/J_1)_r$ in Figs. 14.1 and 14.3, these results imply that one will again see the pendulum phase-space pattern in $(\tilde{\phi}_1, \delta J_1)$ in a plane at fixed ϕ_2, which effectively serves as a clock. This behavior is illustrated in Figs. 14.4 and 14.5. The cross section looks like what we had before in Fig. 13.3.[91] The disrupted region is again of size $\sim \sqrt{\varepsilon}$ around $(J_2/J_1)_r$.

As indicated in Fig. 14.5, this disrupted region will also show up in the nested tori of Fig. 14.3. One way to view the physical phenomenon is with a Poincaré surface of section. Imagine a transverse surface at constant ϕ_2 cutting the tori in Fig. 14.5. Let a particular initial phase point execute its motion.[92] On each cycle, record where it intersects the transverse plane. Repeat for many phase points. This Poincaré surface of section will then display both the undisrupted regions and the disrupted region of size $\sim \sqrt{\varepsilon}$, with the associated islands of stability (and unstable fixed points), as illustrated in Fig. 14.6.

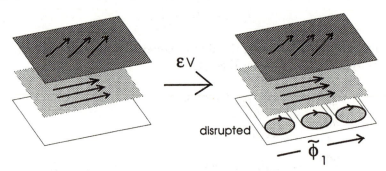

Fig. 14.4. Disrupted region in phase space for two separable periodic systems with an interaction between them of the form in Eq. (14.14); based on Fig. 14.1. Note that $\tilde{\phi}_1 = \phi_1 - (n/m)\,\phi_2$ is now used as abscissa.

A careful reader will note that the model hamiltonian in Eq. (14.14) is symmetric under the interchange $(\phi_1, m) \leftrightarrow (\phi_2, n)$. Thus the canonical transformation in Eq. (14.16) could just as well have been carried out interchanging $(1, m) \leftrightarrow (2, n)$. In either case, one finds resonant disruption of phase space in the single variable $m\phi_1 - n\phi_2$ (as seen in Figs. 14.4–14.6), which involves both ϕ_1 and ϕ_2. In addition, the variables (J_2, ϕ_2) are driven by the coupling to the perturbation δJ_1 [Eqs. (14.25) and (14.27)], and *vice versa*.

---

[91] For the uncoupled case, note that Eqs. (14.3) give $d\tilde{\phi}_1/dt = \omega_1(J_1) - (n/m)\,\omega_2(J_2)$; what are sketched in the disrupted regions in Figs. 14.4 and 14.5 are scaled, stable pendulum orbits in $(\tilde{\phi}_1, \delta J_1)$ as viewed in the $\hat{\phi}_2$ direction (see Fig. 12.2 and footnote 67).

[92] Recall that the motion is 2π-periodic in ϕ_2, so that the nested tori in Fig. 14.5 actually wrap around to repeat when ϕ_2 increases by 2π.

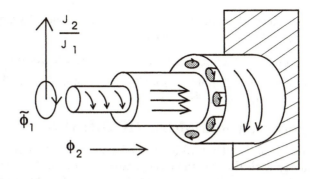

Fig. 14.5. Disrupted region with nested tori and Poincaré surface of section; based on Fig. 14.3. Note that this figure uses $\tilde{\phi}_1 = \phi_1 - (n/m)\,\phi_2$.

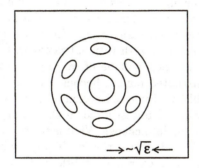

Fig. 14.6. Illustration of intersections of orbits in Poincaré surface of section, with disrupted region, as seen in Fig. 14.5.

It is clear from the present analysis that these results apply to the particular problem of two separable, periodic, weakly coupled systems with the model interaction in Eq. (14.14) of strength ε that retains just one term (m, n) in the perturbation of Eq. (14.9). In this case, it has been demonstrated that one can smooth out the wrinkles in the action-angle phase space $(\tilde{\phi}_1,\ \phi_2,\ J_2/J_1)$ through order ε except for a disrupted torus centered on the resonant value of the ratio of the actions $(J_2/J_1)_r$. This value corresponds to the resonant ratio of angular frequencies

$$\frac{\omega_1(J_{1\,r})}{\omega_2(J_{2\,r})} = \frac{n}{m} \tag{14.29}$$

It was observed that the extent of the disrupted torus is of size $\sim \sqrt{\varepsilon}$ in this phase space.

At least three important issues arise in trying to extend these results:

- What is the effect of working to higher order in ε?

- What is the effect of including more terms in the sum in Eq. (14.9), in particular, an *infinite series* of terms?

- What is the effect of having more than two separable systems?

The celebrated Kolmogorov-Arnold-Moser (KAM) theorem [Ar89a, Sc94, Jo98a, Ot02a] addresses these questions.[93]

The Kolmogorov-Arnold-Moser (KAM) theorem

We are concerned here with weakly coupled separable periodic systems. We have dealt explicitly with the specific case of two systems and worked to lowest order in the coupling strength ε. The condition in Eq. (14.29) states that the disrupted torus occurs when the ratio of angular frequencies is resonant at n/m, the rational number obtained from the active (m, n) term in the double Fourier series expansion of the perturbation in Eq. (14.9). The KAM theorem addresses the effects of keeping the entire series in Eq. (14.9) *and* keeping all orders in ε.

It is evident that one must address the issue of a phase-space disruption at *all possible* resonant (rational) ratios n/m. Even if a particular torus has an irrational ratio, disrupted tori with rational ratios will lie arbitrarily close by. Each isolated disrupted region will be of overall extent $\sim \sqrt{\varepsilon}$. When (m, n) become large, do the disrupted regions become smaller? Do the various regions overlap as $\varepsilon \to 0$? Will the disrupted regions exhibit a fractal structure? Are most of the tori smooth and undisrupted in this limit, or is the whole phase space disrupted and thrown into chaos? What is the total extent of the disrupted region in phase space? It is clear that this is a highly sophisticated mathematical problem, whose resolution is extremely subtle and goes well beyond the level of this supplement. At this point, dedicated readers who wish to pursue the mathematics of this problem are referred to some of the more advanced texts [Ar89a, Li92, Jo98a, Ot02a]. Even a precise mathematical statement of the KAM theorem is inappropriate here. Paraphrased, it asserts that

> *For sufficiently small ε, the sum of all disrupted regions represents a small part of phase space. In the limit $\varepsilon \to 0$, almost all tori (excluding those with rational frequency ratios) are preserved.*

Most irrational tori are displaced and deformed by the small perturbation, but they remain undisrupted.

In connection with the third issue above, it is important to emphasize a fundamental distinction between hamiltonian systems with $N = 2$ degrees of freedom and those with $N \geq 3$. For a hamiltonian system with two degrees of freedom, phase space has four dimensions. Conservation of energy restricts the motion to a three-dimensional subspace, and a given two-dimensional torus divides this space into an inside and an outside. Thus the dynamical motion associated with a disrupted torus remains confined by adjacent undisrupted tori (compare Fig. 14.5) until ε grows large enough to destroy the structure of phase

---

[93]See [Ar89a] for a fascinating historical discussion of the stability of the solar planetary system — a problem that, in part, motivated the theorem.

space. This behavior is especially evident in numerical studies, as seen, for example, in [Jo98a, Ot02a].

The situation is intrinsically more complicated for conservative hamiltonian systems with $N \geq 3$ degrees of freedom [Li92a, Ot02a]. The phase space has dimension $2N$, and energy conservation restricts the motion to a subspace of dimension $2N - 1$. A simple closed "hypersurface" of dimension $2N - 2$ would divide this space into an inside and an outside, but the unperturbed flow of the N angle variables occurs on a torus of dimension N, which is less than $2N - 2$ unless $N = 2$. For any $N \geq 3$, this N-dimensional torus does not divide the energy-conserving subspace (note the analogous situation of a circle in three dimensions, which also fails to divide three-dimensional space). Consequently, for $N \geq 3$, the disrupted chaotic trajectories can extend throughout phase space.

We have generally used a separable hamiltonian as our unperturbed system, which ensures that there enough independent constants of the motion to obtain a complete description of the unperturbed dynamical motion. Such systems are known as "integrable." With increasing perturbation (proportional to ε), however, this simple picture gradually breaks down, as seen in the KAM theorem.[94] More generally, a typical hamiltonian system with N degrees of freedom will have fewer than N independent constants of the motion; such a system is known as "nonintegrable."[95] It is widely believed that nonintegrable systems exhibit chaotic dynamics.[96]

With modern computing capabilities, for example, through Poincaré surfaces of section, it is now possible to study all of these fascinating features in problems of ever increasing complexity. Interested readers are invited to explore for themselves this relatively new and very intriguing field that lies at the interface between mathematics and physics.

---

[94]The success of classical statistical mechanics suggests that with larger ε and very many particles, typical phase-space orbits cover the energy surface sufficiently densely to justify ergodic coarse-grain averaging (see [Wa89, Ot02a]).

[95]For a more thorough treatment of these concepts, see Secs. 7.1.4 and 7.2.1 of [Ot02].

[96]Stone has written a very accessible article on an important early (1917) insight by Einstein concerning the failure of the familiar Sommerfeld-Wilson quantization for nonintegrable systems [St05]. Indeed, Einstein indicates that such nonintegrable systems would be ergodic (although he does not use these precise words).

Part IV
Problems

15 Problems

1.1s[97] Consider the one-dimensional potential

$$V(x) = -\frac{5x}{e^x} + \frac{1}{x^4} + \frac{2}{x} \qquad ; x > 0$$

(a) Plot the potential.

(b) Find and discuss the two points of equilibrium. How many qualitatively different types of motion exist? Find the corresponding phase-space trajectories for different values of the energy and initial conditions. Indicate the energy for each trajectory. Why are these trajectories symmetric with respect to the x axis?

1.2s This problem studies the phase-space orbits for the simple planar pendulum shown in Fig. 1.1.

(a) Derive the hamiltonian $H = (p_\theta^2/2ml^2) + mgl(1 - \cos\theta)$, where the angle θ is measured from the downward vertical.

(b) For small energy $0 < E \ll 2mgl$, verify that the orbits are ellipses.

(c) For $E = 2mgl$, show that the separatrix orbits are given by $p_\theta = \pm 2ml^2\omega_0 \cos\theta/2$, where $\omega_0 = \sqrt{g/l}$.

(d) Use numerical methods to obtain typical phase-space orbits for $E < 2mgl$ (libration) and for $E > 2mgl$ (rotation).

2.1s Equations (2.9) determine the velocity field throughout the fluid for the small-amplitude simple harmonic surface wave studied in the analysis of the Rayleigh-Taylor instability.

(a) Compute and plot this velocity field.

(b) Show that in the first fluid

$$\frac{v_x(x, z, t)}{v_z(x, z, t)} = \coth\left[k(z - h_1)\right] \tan\left[kx - \omega(k)t + \delta\right]$$

where $\Phi_0 \equiv \rho_0 \exp i\delta$. What is the corresponding relation in the second fluid? Discuss.

(c) Show that for this wave the shape of the surface disturbance is given by

$$\zeta(x, t) = \zeta_0 \sin\left[kx - \omega(k)t + \delta\right] \qquad ; \zeta_0 \equiv k\rho_0/\omega(k)$$

2.2s Use Eqs. (2.9) to calculate the pressure everywhere in the fluids for the small-amplitude simple harmonic surface wave studied in the analysis of the Rayleigh-Taylor instability.

2.3s Repeat Prob. 2.1s(a) for the small-amplitude simple harmonic surface wave studied in the analysis of the Kelvin-Helmholtz instability. In this case, the fluids have an unperturbed relative velocity and the velocity potentials are given by Eqs. (2.25).

---

[97] We use the notation s to denote problems associated with the supplement to distinguish them from those in the main text [Fe03].

2.4s Consider the case of a single moving fluid under a vacuum with $\rho_1 = 0$ and $(\rho_2, u_2) \equiv (\rho, u)$.

(a) Show that the dispersion relation in Eq. (2.38) reduces to

$$\bar{\omega}^2 = gk + \frac{\tau k^3}{\rho} \qquad ; \bar{\omega} \equiv \omega - uk$$

(b) Make a galilean transformation to the rest frame of the fluid with a coordinate transformation $\bar{x} = x - ut$. Write $\cos(kx - \omega t) = \cos(k\bar{x} - \bar{\omega}t)$ and identify $\bar{\omega}$ as the angular frequency in the rest frame. Hence show that the result in (a) is just Eq. (2.19) in the limit of a deep tank ($h \to \infty$).

2.5s In a self-gravitating fluid with mass density ρ, the gravitational potential Φ_G obeys Poisson's equation $\nabla^2 \Phi_G = 4\pi G \rho$, where G is the Newtonian gravitational constant. The resulting force per unit mass is $\mathbf{f} = -\nabla \Phi_G$.

(a) Use the equation of continuity, the Euler equation for a compressible nonviscous fluid and Poisson's equation given above to find the dispersion relation $\omega^2 = c^2 k^2 - 4\pi G \rho_0$ for small-amplitude longitudinal plane waves in a uniform self-gravitating medium with mean mass density ρ_0 and speed of sound c (this is a generalization of the treatment in Sec. 49 for sound waves).

(b) What is the physics of this long-wavelength (Jeans) instability for $k < k_J$, where $k_J = \sqrt{4\pi G \rho_0}/c$? The Jeans wavelength is given by $\lambda_J = 2\pi/k_J$ and the Jeans mass is defined as $M_J = \lambda_J^3 \rho_0$. It is believed that the early universe became neutral when the mass density was $\rho_0 \sim 3 \times 10^{-22}$ g/cm^3 and the temperature was $T \sim 3 \times 10^3$ K. Assume $c^2 \sim k_B T/m_p$ and show that $c \sim 5 \times 10^5$ cm/s. Show that $\lambda_J \sim 2 \times 10^{20}$ cm ~ 200 light years and that $M_J \sim 10^6 M_s$, where $M_s \approx 2 \times 10^{33}$ g is the solar mass. The Jeans instability is believed to be associated with the formation of globular clusters, which indeed have these typical masses.

2.6s This problem studies the dispersion relation for transverse waves propagating along a vortex with a hollow core of radius a in an ideal fluid. These "Kelvin" waves are a direct analog of Prob. 10.10, discussed in detail in Sec. 2, with centrifugal force replacing gravity as the restoring mechanism (as in Prob. 10.10, the presence of unperturbed flow makes the analysis somewhat intricate). Kelvin waves have been observed for quantized vortices in both superfluid ^{4}He [Do91] and dilute trapped Bose-Einstein condensates [Br03].

(a) Assume that the unperturbed steady velocity is $v_0(r) = V(r) \hat{\phi}$ for $r \geq a$, where $V(r) = \kappa/(2\pi r)$ and $\kappa > 0$ (see Prob. 9.6). Prove that the flow is both incompressible and irrotational. Find the pressure p_0 throughout the fluid, adjusting the constant so that $p = 0$ at the core. What is the corresponding velocity potential Φ_0?

(b) Study small perturbations of the form

$$\Phi = \Phi_0 + \Phi'$$
$$v = v_0 - \nabla \Phi'$$
$$p = p_0 + p'$$

with the core surface deformed to $r = a + \eta(\phi, z, t)$, where η is a small first-order quantity (here, r, ϕ, z are cylindrical polar coordinates). Obtain the following two relations between η and Φ' evaluated on the unperturbed core:

$$-\frac{V^2(a)}{a}\eta = \left[\frac{\partial \Phi'}{\partial t} + \frac{V(a)}{a}\frac{\partial \Phi'}{\partial \phi}\right]_{r=a}$$

$$\frac{\partial \eta}{\partial t} + \frac{V(a)}{a}\frac{\partial \eta}{\partial \phi} = -\frac{\partial \Phi'}{\partial r}\bigg|_{r=a}$$

(c) Assume $\Phi'(r, \phi, z, t) = \text{Re}\{f(r)\exp[i(m\phi + kz - \omega t)]\}$, where m is an integer. Show that $f(r) = K_{|m|}(kr)$, which is a Bessel function of imaginary argument that vanishes for large values of kr. Combine the various equations to find the dispersion relation

$$\omega_k = \frac{V(a)}{a}\left[m \pm \sqrt{g_m(ka)}\right]$$

where $g_m(x) \equiv -x\,K'_{|m|}(x)/K_{|m|}(x)$ is positive for all integral m.

(d) Take $m = \pm 1$. Explain why the corresponding infinitesimal core deformation $\eta_0 \cos(\pm\phi + kz - \omega t)$ is simply a small displacement of the circular core to a new helical precessing position. For $m = -1$ and $0 < ka \ll 1$, show that there is a low-frequency mode with dispersion relation

$$\omega_k \approx \frac{\kappa k^2}{4\pi}\left[\ln\left(\frac{1}{ka}\right) + c\right]$$

where $c \approx 0.116$ is a constant

Verify that this mode is counter-propagating, which means that it precesses in the clockwise (namely negative) direction at fixed z although the fluid comprising the vortex rotates in the counter-clockwise (namely positive) direction. Show that the corresponding low-frequency mode with $m = 1$ exhibits the same behavior.

3.1s Include the viscous heating term in Eq. (3.1), and state clearly the conditions under which it can be neglected.

3.2s Verify the expression for the static conductive pressure $p^0(z)$ in Eq. (3.15).

3.3s If a fluid rotates with angular velocity Ω, an observer at rest in the rotating frame finds additional noninertial forces (see Prob. 9.27 for the corresponding case of a nonviscous rotating fluid).

(a) In a rotating nearly incompressible viscous fluid, show that the Navier-Stokes equation becomes

$$\frac{\partial \mathbf{v}}{\partial t} + (\mathbf{v} \cdot \nabla)\mathbf{v} = \mathbf{f} - \frac{1}{\rho}\nabla p + \nabla\left(\frac{1}{2}|\mathbf{\Omega} \times \mathbf{r}|^2\right) + \nu\nabla^2\mathbf{v} - 2\mathbf{\Omega} \times \mathbf{v}$$

Discuss the physical interpretation of the rotation-dependent terms.

(b) Show that the first of the linearized Boussinesq Eqs. (3.21) is altered to

$$\frac{\partial \mathbf{v}}{\partial t} = -\boldsymbol{\nabla}\delta w + g\beta \hat{\mathbf{z}}\,\delta T + \nu\nabla^2 \mathbf{v} - 2\boldsymbol{\Omega} \times \mathbf{v}$$

whereas the other two equations are unchanged.

(c) In a rotating fluid, the vorticity $\boldsymbol{\zeta} = \boldsymbol{\nabla} \times \mathbf{v}$ plays an important role. For the Rayleigh-Bénard geometry with $\boldsymbol{\Omega} = \Omega\hat{\mathbf{z}}$, show that the z component of the vorticity ζ_z obeys the boundary condition: $\zeta_z = 0$ at a non-slip surface and $\partial\zeta_z/\partial z = 0$ at a free surface (note that ζ_z vanishes identically for convection rolls in a nonrotating fluid, so that these additional boundary conditions are irrelevant in that case).

4.1s Verify the variational Eqs. (4.19) and (4.20).

4.2s Consider the second unstable mode with Rayleigh eigenvalue $R_2(q)$ in Eq. (4.41). Carry out an analysis similar to that for $R_1(q)$, and make a sketch analogous to that in Fig. 4.1 for the flow pattern of this mode.

4.3s In Sec. 4, we considered the onset of convection for the linearized Rayleigh-Bénard problem with free-free boundary conditions at the top and bottom surfaces. Specifically, we set $\gamma = 0$ and determined the lowest Rayleigh number R that satisfied the Eqs. (4.33).

(a) Now use the more general Eq. (4.30) for the same boundary conditions and the same normal-mode amplitudes $\propto e^{iqx}\sin n\pi z$. Show that the damping constant is given by

$$\gamma(q, n, R, P) = -\frac{Q}{2}\left(1 + \frac{1}{P}\right) \pm \sqrt{\left(\frac{Q}{2}\right)^2\left(1 - \frac{1}{P}\right)^2 + \frac{Rq^2}{QP}}$$

where $Q = n^2\pi^2 + q^2$ and P is the Prandtl number. Find the condition for γ to be positive (namely for the small-amplitude solutions to grow exponentially with time). Show the $n = 1$ is the relevant eigenfunction with $q_c = \pi/\sqrt{2}$ and $R_c = 27\pi^4/4$.

(b) For this case, assume $q = q_c(1 + \delta)$ and $R = R_c(1 + \varepsilon)$ with small δ and ε. Expand γ in this regime to find $\gamma \approx q_c^2(-4\delta^2 + 3\varepsilon)/(1 + P)$. Why can you omit terms of order $\delta\varepsilon$? Discuss the stability and behavior of the conductive heat-flow regime for small negative ε; repeat for small positive ε.

4.4s Verify that with $\gamma = 0$, the x component of Eq. (4.25) is satisfied by the solution found in Eqs. (4.47) and contains no new information. Discuss.

4.5s In a rotating fluid, the velocity and the vorticity are intimately connected. For simplicity, assume that the rotation axis $\boldsymbol{\Omega}$ is along $\hat{\mathbf{z}}$.

(a) Take the curl of the dynamical equation in Prob. 3.3s(b). Use the dimensionless variables from Sec. 3 to obtain the dimensionless equation for the z component of vorticity

$$\frac{\partial\zeta_z}{\partial t} = \nabla^2\zeta_z + 2\tilde{\Omega}\frac{\partial v_z}{\partial z}$$

where $\tilde{\Omega} = \Omega h^2/\nu$ is a dimensionless angular velocity. If $\tilde{\Omega} = 0$, note that the vertical vorticity decouples from the rest of the dynamics; otherwise, v_z and ζ_z are intrinsically coupled.

(b) Take the curl again and obtain the additional dimensionless equation

$$\frac{\partial}{\partial t}\nabla^2 v_z = \nabla^4 v_z + R\nabla_\perp^2 \delta T - 2\tilde{\Omega}\frac{\partial \zeta_z}{\partial z}$$

(c) Assume that the quantities v_z, ζ_z and δT have the time dependence $\propto e^{\gamma t}$ and seek normal modes with the transverse spatial dependence $\propto e^{i\mathbf{q}\cdot\mathbf{r}}$, where $\mathbf{q}$ is in the xy plane. Find the coupled equations analogous to Eqs. (4.30). Analysis of these equations for $\tilde{\Omega} \neq 0$ turns out to be significantly more intricate than for nonrotating fluids (see, for example, [Ch81], Chap. III).

5.1s Derive the vector identities in Eqs. (5.30).

5.2s Verify the last of Eqs. (5.35)

$$(\mathbf{v}\cdot\mathbf{\nabla})\,\zeta(x,z) \;=\; 0$$

used in deriving coupled equations for the leading Fourier amplitudes in the Rayleigh-Bénard problem. Here the velocity is given by Eqs. (5.19) and the vorticity by Eq. (5.28).

5.3s The Nusselt number N is defined in Eq. (3.7).

(a) Use the dimensional form of Eqs. (5.4) and (5.17) to obtain the general result at $z = 0$

$$N = 1 - \sum_{n=1}^{\infty} n\pi\frac{C_{0n}(t)}{\Delta T}$$

(b) Repeat the analysis for $z = h$ and compare [note the discussion following Eq. (5.18)].

6.1s Consider the channel averages introduced in Eqs. (6.4) and (6.5). Discuss how these averages could be made more precise in order to make a quantitative connection between the effective fluid constants $(\mathcal{R}, K)$ and the thermal diffusivity κ and kinematic viscosity ν (see footnote 26).

6.2s Suppose the fluid in Fig. 6.1 is very weakly charged with a uniform charge density $\rho_{ch} = \xi e_0\rho/m$, where ξ is a dimensionless constant and (e_0, m) are appropriate elementary values of charge and mass, The charged fluid is placed in a weak electric field $\mathbf{E} = E_0\hat{z}$. Assume $(e_0, E_0) > 0$ with $\xi e_0 E_0/mg \ll 1$ and work to lowest order in ξ.

a) How are Eqs. (6.1) modified?

b) How are Eqs. (6.22) and (6.25) modified?

7.1s Duffing's one-dimensional oscillator obeys a (dimensionless) nonlinear differential equation

$$\ddot{x}(t) \;=\; -\alpha\,x(t) - \beta\,x^3(t)$$

(a) Find the Lagrangian for this problem.

(b) For $\beta = 0.05$, plot the resulting potential $V(x)$ as a function of α and x in the range $-2 \leq \alpha \leq 2$ and $-10 \leq x \leq 10$. A three-dimensional plot shows how the shape of the potential changes with the parameter α from a single well to a double well. Find the value of α for which this happens (it is called a "bifurcation value" and this particular bifurcation point is known as a "pitchfork").

(c) For $\alpha = -1$, 0, and 1, plot the energy in three dimensions as a function of x and $\dot{x}$. Why does this procedure give the phase-space diagram?

7.2s A van der Pohl oscillator obeys the (dimensionless) nonlinear equation

$$\ddot{x}(t) \;=\; [\, a - x^2(t)\,]\,\dot{x}(t) - x(t)$$

where a is a constant.

(a) By examining various limiting cases, explain qualitatively how the system differs from the damped linear harmonic oscillator.

(b) For $a = 1.0$, integrate the equation numerically for several different initial conditions. For each case, plot $x(t)$ showing several cycles of oscillations and produce the corresponding phase-space trajectory (plot x vs. $\dot{x}$).

(c) How does the attractor differ from that of a damped linear harmonic oscillator?

7.3s For the potential in Eq. (7.4) with initial conditions $x(0) = a$ and $\dot{x}(0) = 0$, use the conservation of energy to find the period $\tau(E)$ as a function of the energy.

(a) How does τ behave as a function of ε for fixed E?

(b) For small ε, expand to rederive the result in Eq. (7.26).

7.4s This problem is preparation for solving more complicated systems of differential equations.

(a) Rewrite Eqs. (23.7) and (23.8) so that they depend only on the normal-mode frequencies ω_1 and ω_2.

(b) For $\omega_1 = 1$ and $\omega_2 = 1.05$, solve the resulting equations numerically for the initial conditions in Eqs. (23.41) and (23.42) with $\alpha = 1$ for some reasonable range of t. Plot the resulting $\eta_1(t)$ and $\eta_2(t)$.

(c) The analytic form of the solutions is given in Eq. (23.49). Plot these solutions for the same values of ω_1 and ω_2 and verify that you have the same behavior to prove that your programs are correct.

(d) For the same frequencies, plot η_2 as a function of η_1. Use a large range of t to see interesting parametric plots (this is a Lissajous figure).

(e) Repeat (b) and (d) for $\omega_2 = 2$ and for $\omega_2 = \sqrt{2}$. Discuss the different behaviors.

8.1s Use the conservation of energy to find the period of the pendulum for both libration and rotation. Compare with the results in Eqs. (8.52) and (8.58). Which approach is simpler?

8.2s Suppose the pendulum hamiltonian is replaced by the following

$$H(p_\theta, \theta) \;=\; \frac{p_\theta^2}{2ml^2} + mgl\left(1 - \cos^2 \frac{\theta}{2}\right)$$

(a) Plot the potential and discuss the motion qualitatively.

(b) Write Hamilton's equations. Find the fixed points and discuss their stability.

(c) Show the action-angle analysis can be recovered from that of the pendulum with the simple replacement $g \to g/2$.

8.3s A particle of mass m moves in a symmetric one-dimensional potential $V(x)$.

(a) Consider a truncated linear potential

$$V(x) = \begin{cases} V_0|x| & \text{for } |x| \leq 1 \\ V_0 & \text{for } |x| > 1 \end{cases}$$

Sketch the potential and identify the regions where periodic motion occurs. For the periodic motions, find the action integral $J = (2\pi)^{-1} \oint p(x)\, dx$. Hence find $E(J)$ and determine the frequency of classical periodic motion as E and J increase.

(b) Use the Sommerfeld-Wilson quantization condition $J = nh$ from Eq. (37.11) to estimate the spacing of quantized energy levels in this potential [this method is especially accurate for large n—the WKB approximation from quantum mechanics [Gr05] shows that an improved approach is to set $J = (n + \frac{1}{2})h$].

(c) Repeat part (a) for the potential $V(x) = V_0\, 2|x|(1 - \frac{1}{2}|x|) = V_0(2|x| - x^2)$. Show that the action integral is

$$J(E) = \frac{\sqrt{2mV_0}}{\pi} \left[\lambda - \frac{1 - \lambda^2}{2} \ln \left(\frac{1 + \lambda}{1 - \lambda} \right) \right]$$

where $\lambda^2 = E/V_0 < 1$. Show that the classical oscillation period diverges like $-\ln(1 - \lambda)$ as $\lambda \to 1$. Compare with the behavior in part (a). Explain why this logarithmic divergence is typical of periodic motion near a separatrix.

8.4s Study the linearized stability near the two qualitatively different kinds of fixed points in Fig. 1.1.

8.5s A driven damped harmonic oscillator obeys the equation

$$\ddot{x} + \sigma \dot{x} + \omega_0^2 x = f_0 \cos(\omega_d t)$$

where ω_0 is the resonant frequency, σ is a damping constant with dimensions of an inverse time, and ω_d is the driving frequency.

(a) Solve for the steady-state solution (ignore the transients that eventually decay away). Discuss briefly how the solution depends on the detuning between ω_0 and ω_d and on the damping σ.

(b) Replace the linear harmonic oscillator with a driven *nonlinear* oscillator that obeys the equation (a driven pendulum)

$$\ddot{x} + \sigma \dot{x} + \omega_0^2 \sin x = f_0 \cos(\omega_d t)$$

For the initial conditions $x(0) = \frac{1}{2}\pi$ and $\dot{x}(0) = 0$, investigate the numerical solution of this equation and compare with the corresponding numerical solution

143

of the linear oscillator. In both cases, plot $x(t)$ and the phase-space trajectory $x(t)$ vs. $\dot{x}(t)$. In particular, consider $\sigma = 0.01$, $f_0 = 0.2$, and $\omega_0 = 1.0$ with $\omega_d = 0.9, 1.0, 1.1$. What happens if $\omega_d = 0.5$ and $\omega_d = 2.0$ (along with $\omega_d = 1$, these are examples of resonant driving forces)? What happens for an irrational ratio such as $\omega_d = 1/\sqrt{2}$? What happens for larger driving force f_0?

8.6s Consider the (dimensionless) nonlinear second-order differential equation

$$\ddot{x} = \dot{x}^3 - x^3 + x$$

Convert into two coupled first-order equations, and then
    (a) Find the fixed points.
    (b) Show that the fixed point at the origin is unstable, while the other two fixed points are stable, but not attracting.

9.1s Derive the critical value $r^* = P(P+b+3)/(P-b-1)$ given below Fig. 9.1 for the onset of oscillatory growth (namely where $\text{Re}\lambda$ first becomes positive).

9.2s Write, or obtain, a program to solve the nonlinear coupled Lorenz Eqs. (9.3). Use $r = 28$ and the starting values $(x_0, y_0, z_0) = (10, 10, 27)$.
    (a) Reproduce the result in Fig. 9.2.
    (b) Carry the calculation out as far as you can in t.
    (c) Determine when the reproducibility of your result is limited by the tolerances in your calculation (error in the integration algorithm, computer round-off error, *etc.*).

9.3s Use the program of Prob. 9.2s to extend the results in Fig. 9.6 to higher t, and observe the continued shrinking of this phase space.

9.4s In discussing the periodic orbit arising from the solution to the Lorenz Eqs. (9.3) at $r = 148.5$ it is stated that, "If the value of r is decreased slightly to $r = 147.5$, close inspection shows that the orbit is actually a union of two congruent orbits slightly separated in (x, z) space, and it takes just twice as long for the system to complete one full orbit — the *period is doubled*." Can you numerically verify this statement?

9.5s Repeat the numerical analysis of the Lorenz equations for $P = 3$ and $b = 1$, and confirm that the behavior for various r is essentially the same as that found in Sec. 9 for $P = 10$ and $b = 8/3$. What happens if $P < 2$ and $b = 1$?

9.6s Rewrite the equations for the van der Pohl oscillator (Prob. 7.2s) as a pair of coupled first-order nonlinear ordinary differential equations.
    (a) Is this a hamiltonian system? How does the volume in phase space change with time? What happens as $t \to \infty$?
    (b) Repeat for the Duffing oscillator (Prob. 7.1s).

10.1s Section 10 relies largely on analytic and numerical methods. For an illuminating graphical analysis of the logistic map, see [Jo98b, Ot02b]. Write an account of this graphical approach to the first two period doublings at $\rho = 3$ and $\rho = 3.449\cdots$.

10.2s Consider the logistic map defined in Eqs. (10.9).

(a) Write, or obtain, a program to iterate this finite difference equation.

(b) Reproduce Figs. 10.1–10.2(a), and extend them to higher n. Show that the oscillations are stable against the starting value of x_0.

(c) Can you identify an eight-cycle oscillation for $\rho > 3.54$?

10.3s Consider the double bifurcation into a four-cycle oscillation of the logistic map as illustrated in Fig. 10.5.

(a) Find a numerical method to determine that this occurs at $\rho = 3.449\cdots$.

(b) Determine x_{min} and x_{max} at this value of ρ.

10.4s Carry out a linear stability analysis about the new fixed points to show that the two-cycle oscillation in Fig. 10.5 is stable for $3 < \rho < 1 + \sqrt{6} = 3.4494\cdots$.

11.1s Consider the free one-dimensional motion of a mass point.

(a) Show that the phase orbit which originates at (p, q) at $t = t_0$ is given parametrically by

$$P = p(t) = p$$
$$Q = q(t) = q + \frac{p}{m}(t - t_0)$$

(b) Show the jacobian determinant of the transformation from (p, q) to (P, Q) is unity for all t.

(c) Consider the time development of a finite right rectangle at $t = t_0$ in (p, q) phase space. Show that although the shape of this region changes with time, the phase area is preserved.

11.2s Consider the one-dimensional simple harmonic oscillator of Eqs. (8.2).

(a) Show that the phase orbit which originates at (p, q) at $t = t_0$ is given parametrically by

$$P = p(t) = p \cos\alpha - m\omega_0 q \sin\alpha$$
$$Q = q(t) = \frac{p}{m\omega_0}\sin\alpha + q\cos\alpha \qquad ; \alpha \equiv \omega_0(t - t_0)$$

(b) Show the jacobian determinant of the transformation from (p, q) to (P, Q) is unity for all t.

(c) Consider the time development of a finite area at $t = t_0$ in (p, q) phase space. Use the above equations to conclude that this area is unchanged.

12.1s Consider a system of two uncoupled pendulums. Write the analogs of Eqs. (12.17) to (12.26) for the hamiltonian and action-angle analysis of this system.

12.2s Consider the two uncoupled simple harmonic oscillators of Eqs. (12.32).

(a) Calculate and plot a few illustrative toroidal orbits in the three-dimensional coordinates of Fig. 12.2.

(b) Make the two-dimensional plots corresponding to Fig. 12.3.

13.1s Rewrite Eqs. (13.29) in the dimensionless variables of Eq. (7.27)

$$\dot{z} = -4\lambda z^2 \sin^3 \phi \cos \phi$$
$$\dot{\phi} = 1 + 2\lambda z \sin^4 \phi$$

where $x = \omega_0 t$, $\lambda = \varepsilon a^2/\omega_0^2$, and $z = J/ma^2\omega_0$ with a the maximum displacement of the unperturbed oscillator. Here the dot indicates a "time" derivative with respect to x.

(a) Write, or obtain, a program to solve these coupled nonlinear first-order differential equations.

(b) Verify the rippled structure sketched in Fig. 13.1 for small values of λ.

(c) Discuss what happens for larger λ.

13.2s Generalize the treatment of the anharmonic oscillator in Eq. (13.25) to the perturbation $\varepsilon|q|^n/n$.

(a) Express the hamiltonian in action-angle variables and evaluate the mean hamiltonian $\bar{H}(J) = \langle H \rangle$ averaged over the range $0 \leq \phi \leq 2\pi$. Use $\bar{H}$ to obtain the perturbed frequency

$$\omega = \omega_0 + \frac{\varepsilon}{m}\left(\frac{2J}{m\omega_0}\right)^{n/2-1} \frac{\Gamma(n/2 + 1/2)}{\sqrt{\pi}\,\Gamma(n/2 + 1)}$$

(b) Reproduce the previous result for $n = 4$ and find the corresponding shift for $n = 3$.

(c) Obtain the same results with the method used to obtain Eq. (7.26). Which approach is simpler?

14.1s Generalize the treatment of Eq. (13.28) to the case of two coupled oscillators with

$$H = \frac{p_1^2}{2m} + \frac{1}{2}m\omega_1^2 q_1^2 + \frac{p_2^2}{2m} + \frac{1}{2}m\omega_2^2 q_2^2 + \varepsilon m^2 \omega_1^2 \omega_2^2 q_1^2 q_2^2 \tag{15.1}$$

Transform to action-angle variables $(J_1, J_2, \phi_1, \phi_2)$, and construct the new hamiltonian $H(J_1, J_2, \phi_1, \phi_2)$. Use the mean hamiltonian $\bar{H}(J_1, J_2) = \langle H \rangle$ averaged over the range $0 \leq \phi_1, \phi_2 \leq 2\pi$ to show that the perturbed frequencies are $\approx \omega_1 + \varepsilon\omega_1\omega_2 J_2$ and $\approx \omega_2 + \varepsilon\omega_1\omega_2 J_1$ correct to first order in ε.

14.2s Consider the hamiltonian $H(J_1, J_2, \phi_1, \phi_2)$ of Prob. 14.1s.

(a) Write Hamilton's equations for the variables $(J_1, J_2, \phi_1, \phi_2)$. Make them dimensionless.

(b) Write, or obtain, a program to solve these four coupled nonlinear equations.

(c) Construct the Poincaré surface of section for given values $(E, \phi_2) = (E, \phi_2)_{\text{fixed}}$ for some set of dynamical trajectories. (*Hint*: look for trajectories in the subspace $E = E_{\text{fixed}}$.) Discuss.

References

[Ab64] M. Abramowitz and I. A. Stegun (eds), *Handbook of Mathematical Functions with Formulas, Graphs, and Mathematical Tables*, National Bureau of Standards, *Applied Mathematics Series* **55**, U. S. Government Printing Office, Washington, D. C. (1964), Chap. 17

[Am03] P. Amore and A. Aranda, *Phys. Lett.* **A316**, 218 (2003)

[Ar80] V. I. Arnold, *Ordinary Differential Equations*, MIT Press, Cambridge, MA (1980)

[Ar89] V. I. Arnold, *Mathematical Methods of Classical Mechanics, 2nd ed.*, Springer-Verlag (1989)

[Ar89a] *ibid.* Appendix 8

[Be78] C. M. Bender and S. A. Orszag, *Advanced Mathematical Methods for Scientists and Engineers*, McGraw-Hill, New York (1978), p. 352

[Be78a] *ibid.* Sec. 11.1

[Bo00] E. Bodenshatz, W. Pesch, and G. Ahlers, *Ann. Rev. Fluid Mech.* **32**, 709 (2000)

[Br03] V. Bretin, P. Rosenbusch, F. Chevy, G. V. Shlyapnikov, and J. Dalibard, *Phys. Rev. Lett.* **90**, 100403 (2003)

[Ch79] B. V. Chirikov, *Phys. Rept.* **52**, 265 (1979)

[Ch81] S. Chandrasekhar, *Hydrodynamic and Hydromagnetic Stability*, reprinted by Dover Publications, Mineola, New York (1981)

[Ch81a] *ibid.* Chap. X

[Ch81b] *ibid.* Chap. XI, Sec. 101

[Ch81c] *ibid.* Chap. II

[Cr93] M. C. Cross and P. C. Hohenberg, *Rev. Mod. Phys.* **93**, 851 (1993)

[Cv84] P. Cvitanović, *Universality in Chaos: A Reprint Selection*, Adam Hilger, Bristol (1984)

[Do91] R. J. Donnelly, *Quantized Vortices in Helium II*, Cambridge University Press (1991), Chap. 6

[Fe78] M. J. Feigenbaum, *J. Stat. Phys.* **19**, 25 (1978)

[Fe80] M. J. Feigenbaum, *Los Alamos Science* **1**, 4 (1980), reprinted in [Cv84]

[Fe80a] A. L. Fetter and J. D. Walecka, *Theoretical Mechanics of Particles and Continua*, McGraw-Hill, New York (1980)

[Fe80b] *ibid.* Problems 10.8, 10.9, 10.10

[Fe03] A. L. Fetter and J. D. Walecka, *Theoretical Mechanics of Particles and Continua*, reprinted by Dover Publications, Mineola, New York (2003)

[Gr05] D. J. Griffiths, *Introduction to Quantum Mechanics, 2nd ed.*, Pearson Prentice Hall, Upper Saddle River, NJ (2005), Chap. 8

[Gu90] M. C. Gutzwiller, *Chaos in Classical and Quantum Mechanics*, Springer-Verlag (1990)

[Gu90a] *ibid.* Chap. 9

[Hi74] M. W. Hirsch and S. Smale, *Differential Equations, Dynamical Systems, and Linear Algebra*, Academic Press, New York (1974)

[Jo98] J. V. José and E. V. Saletan, *Classical Dynamics: A Contemporary Approach*, Cambridge University Press (1998)

[Jo98a] *ibid.* Sec. 7.5

[Jo98b] *ibid.* Sec. 7.4

[Ka83] L. P. Kadanoff, *Road to Chaos*, in *Physics Today*, December (1983), p. 46

[La60] L. D. Landau and E. M. Lifshitz, *Mechanics*, Pergamon, Oxford (1960), Sec. 46

[La80] L. D. Landau and E. M. Lifshitz, *Statistical Physics, 3rd ed., Part 1*, Pergamon, Oxford (1980), Secs. 142-143

[La87] L. D. Landau and E. M. Lifshitz, *Fluid Mechanics, 2nd ed.*, Pergamon, Oxford (1987)

[La87a] *ibid.* Sec. 56

[La87b] *ibid.* Sec. 57

[Li92] A. J. Lichtenberg and M. A. Lieberman, *Regular and Chaotic Dynamics, 2nd ed.*, Springer-Verlag (1992)

[Li92a] *ibid.* Chap. 6

[Lo63] E. N. Lorenz, *J. Atmos. Sci.* **20**, 130 (1963), reprinted in [Cv84]

[Ma76] R. M. May, *Nature* **261**, 459 (1976), reprinted in [Cv84]

[Mc75] J. B. McLaughlin and P. C. Martin, *Phys. Rev.* **A12**, 186 (1975)

[Mc94] J. L. McCauley, *Chaos, Dynamics, and Fractals: an Algorithmic Approach to Deterministic Chaos*, Cambridge University Press (1994)

[Ot02]   E. Ott, *Chaos in Dynamical Systems, 2nd ed.*, Cambridge University Press (2002)

[Ot02a] *ibid.* Secs. 7.2-7.4

[Ot02b] *ibid.* Sec. 2.2

[Ot02c] *ibid.* Sec. 4.4

[Pe92]   I. Percival and D. Richards, *Introduction to Dynamics*, Cambridge University Press (1992)

[Pe99]   M. Peskin, edited and with additional material by S. Doniach, *Lecture Notes on Advanced Particle Mechanics*, Stanford (1999), unpublished

[Ru86]   R. D. Ruth, *Single Particle Dynamics in Circular Accelerators*, Invited talk presented at the 1985 SLAC Summer School on the Physics of High Energy Particle Accelerators, Stanford, CA, July 15-26, 1985, SLAC-Pub-4103, October 1986

[Sa62]   B. Saltzman, *J. Atmos. Sci.* **19**, 329 (1962)

[Sc94]   F. Scheck, *Mechanics, 2nd ed.*, Springer-Verlag (1994), Secs. 2.38-2.39

[Sp82]   C. Sparrow, *Simple Properties of the Lorenz Equations*, in *Appl. Math. Sci.* **41**, Springer-Verlag (1982), p. 1

[St05]   A. D. Stone, *Einstein's Unknown Insight and the Problem of Quantizing Chaos*, in *Physics Today*, August (2005), p. 37

[Wa89]   *Fundamentals of Statistical Mechanics: Manuscript and Notes of Felix Bloch, Prepared by J. D. Walecka*, Stanford University Press, Stanford, California (1989); reissued by World Scientific Publishing Company, Singapore (2000)

[Wa04]   J. D. Walecka, *Theoretical Nuclear and Subnuclear Physics, 2nd ed.*, World Scientific Publishing Company, Singapore (2004)

[Wo96]   S. Wolfram, *The Mathematica Book, 3rd ed.*, Wolfram Research, Inc., Champaign, Illinois (1996), pp. 750 and 1074

[Wo05]   S. Wolfram, *Elliptic Integrals and Elliptic Functions*, Mathematica web site http://functions.wolfram.com

[Yo85]   J. A. Yorke and E. D. Yorke, *Chaotic Behavior and Fluid Dynamics*, in *Topics in Applied Physics* **45**, eds. H. C. Swinney and J. P. Golub, Springer-Verlag (1985), p. 77

[Za85]   G. M. Zaslavsky, *Chaos in Dynamical Systems*, Harwood Academic Publishers, New York (1985)

Index